John Ziman is one of the most influential writers on the practice of science. In the course of a distinguished career in the natural sciences he became concerned about the social relations of science and technology. He was for many years chairman of the Council for Science and Society, and between 1986 and 1991 he headed the Science Policy Support Group. He is as much at ease writing a specialist research paper as a newspaper article. His books include: *Reliable Knowledge, Teaching and Learning about Science and Society, An Introduction to Science Studies* and *Knowing Everything about Nothing*, all published by Cambridge University Press.

After expanding for centuries, science is reaching its limits to growth. We can no longer afford the ever-increasing cost of exploring ever-wider research opportunities. In the competition for resources, science is becoming much more tightly organized. A radical, pervasive and permanent structural change is taking place. This already affects the whole research system, from everyday laboratory life to the national budget. The scientific enterprise cannot avoid fundamental change, but excessive managerial insistence on accountability, evaluation, 'priority setting', etc. can be very inhospitable to expertise, innovation, criticism and creativity. Can the research system be reshaped without losing many features that have made science so productive? This trenchant analysis of a deep-rooted historical process does not assume any technical knowledge of the natural sciences, or their history, philosophy, sociology, or politics. It is addressed to everybody who is concerned about the future of science and its place in society.

Related title

Reliable Knowledge: An exploration of the grounds for belief in science
JOHN ZIMAN

JOHN ZIMAN

Prometheus Bound

Science in a dynamic steady state

The Father of the Gods wished to punish the Titan who had stolen Fire from Heaven and brought it to the People of Earth. Prometheus was chained to a mountain top, and savaged day after day by a voracious eagle.

CAMBRIDGE UNIVERSITY PRESS

Published by the Press Syndicate of the University of Cambridge
The Pitt Building, Trumpington Street, Cambridge CB2 1RP
40 West 20th Street, New York, NY 10011-4211, USA
10 Stamford Road, Oakleigh, Melbourne 3166, Australia

First published 1994

Printed in Great Britain at the University Press, Cambridge

A catalogue record for this book is available from the British Library

Library of Congress cataloguing in publication data
Ziman, J. M. (John M.), 1925–
Prometheus bound : science in a dynamic "steady state" / John Ziman.
p. cm.
Includes index.
ISBN 0-521-43430-0
1. Science and state. 2. Science—Social aspects. 3. Research—
Management. I. Title.
Q125.Z56 1994
338.9'26—dc20   93-5922   CIP

ISBN 0 521 43430 0 hardback

Contents

Preface

Science is reaching its 'limits to growth'. It is expected to contribute increasingly to national prosperity, yet national budgets can no longer support further expansion to explore tempting new research opportunities, by larger research teams, equipped with increasingly sophisticated apparatus. As a result, science is going through a radical structural transition to a much more tightly organized, rationalized and managed social institution. Knowledge-creation, the acme of individual enterprise, is being collectivized.

This transition is pervasive, interlocking, ubiquitous and permanent. It affects the whole research system, from the everyday details of laboratory life to the politics of national budgets. Changes in one part of the system, such as the abolition of academic tenure, have repercussions elsewhere, for example in the commercial exploitation of scientific discoveries. A new policy language of 'accountability', 'evaluation', 'input and output indicators', 'priority-setting', 'selectivity', 'critical mass', etc. has become commonplace throughout the world, from Finland to Brazil, from Poland to New Zealand, from the United States to Papua New Guinea. Indeed, science is becoming a truly international enterprise, organized systematically on a global scale.

Many scientists and scholars look back regretfully to a more relaxed and spacious environment for academic research. But nostalgia is a fruitless sentiment. What all scientists know is that science cannot thrive without social space for personal initiative and creativity, time for ideas to grow to maturity, openness to debate and criticism, hospitality towards innovation, and respect for specialized expertise. The real question is not whether the structural transition is desirable, or could have been avoided: it is how to reshape the research system to fit a new environment without losing the features that have made it so productive in the past.

Although this question is unanswerable, it informs action merely by being asked. But that means that it must be clearly posed. The general idea of a major structural change has been around for a long time. It underlies the eloquent writings of Derek de Solla Price (1963), Jean-Jacques Salomon (1970) and Jerome Ravetz (1971). Since then, working scientists in the UK and elsewhere have become uncomfortably accustomed to level funding, and complain loudly about its various effects on the research enterprise.

Nevertheless, this striking phenomenon has not yet become a major research site in the world of science and technology studies. I first began to formulate the main issue myself in a short article (*Minerva*, **16**, 327–39, 1978) entitled 'Bounded science: the prospect of a steady state' – actually a review of Bruce Smith and Joseph Karlesky's disturbing account of the then state of academic science in the USA. I also began to notice various structural effects, such as increasing societal influence on problem choice in fundamental research (*Minerva*, **19**, 1–42, 1981), the shift away from the traditional individualism of scientific activity (Bernal Memorial Lecture, *Proc. R. Soc. Lond. B*, **219**, 1–19, 1983), and changing career patterns in basic science (*Knowing everything about nothing*, 1987).

These strands were eventually woven together in 1987, in a series of informal workshops organized by the Science Policy Support Group (SPSG). About a dozen very senior people, from all sectors of the UK research system, took part in these discussions; Science in a 'Steady State': The research system in transition (SPSG Concept Paper No.1, December 1987) sketched out the remarkably coherent picture that emerged from the convergence of their experience and opinions. The present book is essentially an enlargement of this picture, clarified and made more vivid for a wider circle.

In writing it, I had in mind the 'attentive public' for science, typified perhaps by regular readers of *New Scientist*, or regular watchers of *Horizon* or *Equinox*. I have tried to use direct, 'quality newspaper' language, free of the scholarly impedimenta of foot-notes and bibliographic references. I have not assumed any tech-nical knowledge of the natural sciences, their history, philosophy, sociology, or politics. The argument is illustrated by numerous particular examples, mostly drawn from the countries of the English-speaking world, but reference is made to parallel devel-opments in other countries, especially in Europe. A neutral ideo-logical stance on political and economic issues is attempted, although my personal preferences (if not prejudices) are not very carefully concealed. In particular an intemperate and ungainly deployment of shriek quotes – e.g. 'accountability' or 'selectiv-ity' – clearly indicates my deep misgivings about the way that perfectly reasonable words have been degraded into fashionable terms of art for some very questionable practices.

For good or ill, this text, as it now stands, is entirely my own responsibility. But a work such as this owes a great deal to the friendly interest of many well-informed people. A full list of these would be very long indeed, and a short list would be invidious. It would include, first of all, the original SPSG working party, who actively commented on successive drafts of the SPSG Con-cept Paper. Add to that the hundred participants in the two-week NATO Advanced Study Institute at Il Ciocco, Italy, in October 1989, where the theme of 'Managing Science in the Steady State' was analysed from a number of different points of view (see Cozzens *et al.*, 1990). In addition, my thinking on the subject has benefited greatly from the penetrating questioning of many alert scientists and metascientists to whom I have presented the argu-ment at conferences and seminars in a number of different coun-tries. Even if I don't attempt to name you all, be fully assured of my gratitude. This is also an ideal opportunity to acknowledge the personal pleasure and profit of five years of fruitful cooperation with my former colleagues in the Science Policy Support Group – Peter Healey, Lynda Robb, Carlye Honig, Lisa Harding and Michael Hanna – who provided the organizational base that made this project possible.

John Ziman
Oakley, Bucks.

1
What is happening to science?

The dark cavern in the mountain rang with the merry hammering of the Dwarves.

1.1 The shape of science today

Imagine a space–time traveller, returning after 30 of our earthly years. For Dr Winkel van Rip, it seems only a few months since she hurtled off at nearly the velocity of light in a secret biotest of General Relativity. Now she has come back to her job as lecturer in astrophysics at Loamshire University. Until now she has been a dedicated researcher, entirely uninterested in any aspect of science beyond her textbooks and laboratory bench. What features of contemporary science might catch her innocent eye? What account would she give of scientific life and work in Britain today?

At first, she would be delighted and overwhelmed by all the good science that had been done in her absence. She would enjoy learning about the theoretical explanations of old mysteries, and getting her hands on to marvellously powerful new observational techniques. She would also soon realise that many of the old questions had still not been answered, and that many exciting new questions were emerging. Astrophysics happens to have made conspicuous progress in the last thirty years, but it is not unique. In almost every other field of the natural sciences she would find the same buzz of activity, immense technical achievements, and ever widening challenges and opportunities.

She would also note that science and its technological products was highly prized in society generally, especially as a source of national economic achievement. She would see scientists using incredibly refined techniques to acquire quite remarkably powerful understanding of every aspect of the natural world. She might feel slightly uneasy at the attention given to the voices of intellectual dissent, and to ethical concerns about the effects of technological innovation. Nevertheless, her 1960s vision of unabated and beneficial scientific progress would still seem to be alive and well in the 1990s.

But as she began to work again at her profession she would begin to realize that the whole scene had subtly changed. Being a wise person, she would not take too seriously the perennial grumbles about lack of research funds, inept decisions by ill-informed bureaucrats, political incomprehension of the value of science to the nation, etc. Scientists have always felt inadequately appreciated by those who do not share their peculiar, esoteric missions and have always complained that they deserved far more resources and social recognition. She might have noted, however, that these discontents were now being articulated more precisely, and backed by facts and figures. Responsible people were questioning the future of British science, if not of science as a whole. Even in their table talk, scientists would seem to be much more politically anxious and self-conscious than they used to be.

Moreover, in listening to this table talk, the good Dr van Rip would frequently hear words that seemed quite out of place in the academic world – words such as 'management', 'outputs', 'accountability', and 'evaluation'. Piecing these buzzwords together, and puzzling out the meaning of various unfamiliar terms such as 'critical mass' and 'intellectual property rights', she would begin to see the science of today in quite a new light.

First of all, and most importantly, she would find herself involved in a single, loosely articulated but essentially indivisible, **'R&D' (research and development) system**, where very diverse activities, ranging from basic scientific research to near-market technological development, interpenetrate and combine within a bewildering variety of institutions. She would note references to an elaborate apparatus of committees and administrators designed to provide a **policy** for this system. This policy would include such matters as: formulating national **priorities** in deciding how research should be funded; achieving early identification of **exploitable** areas of academic science; increasing the emphasis on **strategic** and **pre-competitive** research; initiating national programmes of **directed** research in fields such as information technology and biotechnology; fostering direct *industrial/academic* linkages; negotiating transnational programmes of research and technological development through international organizations such as the European Community.

Her academic colleagues would provide her with a running commentary on some of the characteristics of the national machinery for the allocation of **resources**. They would tell her about the fierce **competition** between scientists to get funds for research, the insistence on increased **accountability** for the inputs and outputs of science, and on systematic accountancy for the **overheads** incurred in carrying out research. She would also learn about the politics of setting up and arranging the shared use of **facilities**, often on a multinational basis, such as particle accelerators, synchrotron radiation sources, computer networks, etc.

Our innocent friend would also note the development of all sorts of procedures for the systematic evaluation of the performance of every element of the research system: research units, laboratories, research establishments, university departments, Higher Educational Institutions of various kinds, and even whole nations. She would observe that the work of **individual** academic scientists was subject to periodic **appraisal**, and that various performance **indicators** had been developed to monitor **efficiency** and facilitate **selectivity** in the allocation of resources according to presumed quality. Although the term triage would not be used publicly, she would become aware of various administrative devices for killing off weak research units to enhance the survival of the others.

In the same context, she would hear of various measures designed to concentrate scientific activities, in order to achieve a **critical mass** of effort. These measures might include the establishment of national **centres of excellence** designed to compete internationally in particular subjects, the agglomeration of scientists and equipment into **multidisciplinary** centres to attack particular practical problems, and the development of research **networks** linking scientists in different institutions.

In considering her own further career prospects as a scientist, she would realise with alarm that she could no longer rely upon permanent tenure in her university post, and might find herself among the ranks of **contract researchers** on short-term appointments supported by **soft money**. She would also become conscious of a number of other features of the system, including concern about the efficiency of professional research **training**, a trend towards the separation of **teaching** from research, and the association of **managerial** functions with scientific leadership roles.

Forgive me for the above paragraphs. They lampoon policy prose to make a serious point. Of course, the table talk would mostly be taken up with other, more familiar topics. Within a few weeks, for sure, our time traveller would find herself entirely at home again in the interminable gossip of a peculiarly individualistic profession. She would soon be discussing the merits and demerits of the latest professorial appointment, the rights and wrongs of a notoriously critical referee's report, the slightly dubious circumstances surrounding the decision of Dr A to go to America, the advantages and disadvantages of collaborating with the rather snooty group at X, the choice of invited speakers for the next international conference, the selection of suitable themes for a forthcoming Summer School, or the prospects for next year's Nobel Prize. Most of the strong and abiding traditions of the scientific life would still seem reassuringly alive. But was this perhaps a superficial view? How was the whole world of science and scholarship being affected by the practices, procedures and attitudes signalled by those peculiar words that kept buzzing into the conversation?

1.2 Radical, irreversible, structural change

Like most of her fellow scientists, Dr van Rip would not have a deep historical perspective on the scientific enterprise as a whole. She would probably have some difficulty in fitting her findings together into a coherent picture of that enterprise today. But being very observant, she would note certain common characteristics of the changes that had taken place since she had started on her journey.

In the first place, the novel features are *widespread and penetrating*. Within a few weeks of entering the scientific world, at any point, our observer would have encountered most of them. They are not just the commonplaces of coffee-break chat, or the subject matter of government reports, or editorials in *Nature* or *New Scientist*. They affect the everyday decisions and working practices of researchers in universities, government establishments and industrial laboratories. They motivate committee agendas and give professors ulcers. They are deeply embedded in the *culture* of research: they form the *climate* in which scientists have to live.

They have *many different aspects*. Each of those cryptic phrases, tagged with its buzzword, stands for an ill-defined complex of procedures and principles, policies and practices. Take, for example, the requirement that the **overheads** of academic research must be accounted for. This obviously includes expendable items such as laboratory reagents, or the materials used in making research apparatus in the departmental workshop. It has to cover secretarial facilities and libraries; but what about the central university administration? At what rate should expensive scientific instruments be written off for obsolescence? How much should be set aside for the putative rent of that grand new, or decrepit old building? It looks like a standard exercise in accountancy, but conceals an infinite source of unproductive institutional toil and moil.

The novel features are *interconnected in many different ways*. Take again the apparently innocuous theme of **overheads**. The whole calculation depends on the degree to which **teaching** is separated from research. It is affected by economies of scale achieved by the **concentration** of resources into larger units, and by the **networking** of facilities. It must be a consideration in the **competition** between research groups, and in their relative **evaluation**. And so on. These features are so interlaced that every time she tried to break the list down into separate categories she would have come out with a different scheme!

The same changes would be seen to have occurred, to a greater or lesser degree, in *most other scientifically developed countries*. The verification of this proposition would be a major task, beyond her personal capacity. But if she were to test her list on an audience of senior scientists and policymakers from most of Western Europe, North America and Australasia, she would not be seriously contradicted. The countries of Central and Eastern Europe, including the successor states to the Soviet Union, are working hard to restructure their science along similar lines. Communications from OECD (Organization for Economic Cooperation and Development) and UNESCO (United Nations Educational, Scientific and Cultural Organization) suggest that many of the Less Developed Countries are moving, or are being pushed by their advisers, in the same direction. In Japan, perhaps, the situation is rather different, although not at certain significant points such as the focus on **exploitation**.

Finally, our quantum leaper would sense that most people now believed that these new features were *unlikely to disappear in the foreseeable future*. However regrettable some of them may seem to many of the people affected by them, the general thrust of communal effort, from ginger groups as much as from establishment circles, is to accept them as permanent features of scientific activity and do their best to 'make them work', and 'get them right'. In many countries, particularly the UK, any call for these changes to be reversed is swamped by the demand for additional resources. Science everywhere is agonised by a thirst for money, both for positive new developments and to meet the cost of installing all these features into its existing institutions.

Now let Dr van Rip cast her mind back to the time of her apprenticeship to research – little more than a human generation ago. Let her refresh her memory of the scientific life in a country such as Britain in the early 1960s. Let her read the autobiographies, listen to the reminiscences, and skim through old professional journals, newspapers and documents. She would surely be nonplussed by the realisation that *most of the features that are now so dominant would then have been scarcely noticeable.*

A serious historical scholar could, of course trace some of the basic themes, such as public pronouncements on the desirability of links between industry and academia, back to Francis Bacon. 'Evaluation' and 'selectivity', in one form or another, have been key factors in the building of scientific institutions and scientific careers since research went professional in the nineteenth century. But they were not highlighted and trumpeted around as indispensable tools of 'management' – a word that would never have been heard in academic circles. And other present-day features, such as job appraisal and abolition of tenure for university staffs, would have been quite inconceivable at that time.

It is not necessary to give a detailed historical account of each one of all the novel features that have emerged or evolved into prominence in this short period. They are so numerous, so pervasive, and so interconnected, that they amount to convincing empirical evidence of a profound structural change. In less than a generation we have witnessed a *radical, irreversible, world-wide transformation* in the way that science is organized and performed.

1.3 A provisional analysis for practical guidance

This, in essence, is what an objective observer of the present-day scientific world might report. But those of us who are immersed in this world find it difficult to appreciate the magnitude of the transformation that has occurred. On the whole, it has been a continuous process, leading to a situation to which the older generation have adapted reluctantly and which the younger generation accept as normal. We are all aware that many significant changes have taken place, but we have tended to treat these as independent phenomena, generated almost accidentally by peculiar political or economic circumstances beyond our control. Until quite recently, very little effort had been made to connect them up and locate them in a more general scheme.

What is it all about? Why has it happened? If this were a scholarly text in an academic course of science studies, there would now have to be a systematic analysis of the phenomena in question. Although there seems little doubt that a very significant change in the social structure of science has indeed taken place, a proper historical account would probably not make it seem quite so dramatic or abrupt as I have here suggested. There would be references to trends that were already under way between the World Wars, to the prophetic writings of J.D.Bernal and others, and to the profound effects of war-time developments on the scale and influence of the sciences and their associated technologies. Many items in the current vocabulary of science policy were being formulated in the 1950s by Alexander King, from his post in OECD, and so on.

My guess is, moreover, that the concept of a 'new regime' or a 'new model' for science was already in the air towards the end of the 1960s, even though the big changes did not begin to take place on the ground until about a decade later. I am quite deliberately evading any question about the nature of this consciousness or how it developed. Jerry Ravetz' notion of 'industrialization' and Jean-Jacques Salomon's concept of 'technoscience' come to mind, as well the sharply contrasting works on the 'politics' of science by Don Price and Daniel Greenberg; but we still lack a serious historical study of the process of change.

What we also lack is an agreed sociological account of science under these new conditions. Arie Rip (1989) and others are working on this, but it is obviously extremely complex, and likely to become very controversial. The transition tends to be treated from an 'externalist' viewpoint, as if all that had happened was an intensification of the political, economic, military and industrial forces to which science has – on this view – always been highly responsive. In my opinion, there is a lot of work to be done in developing an alternative 'internalist' account, in which a relatively self-contained institution would be seen as going through a traumatic cultural change under the impact of these forces. But these are issues for the scholars, who are invited to dispute them as they will.

The purpose of the present book is much less exalted. The scientific world is in turmoil. Working scientists, scientific leaders, industrial managers, government officials, academic authorities, and many others are seeking guidance on how to carry out their duties in a rapidly changing landscape of institutions and procedures. The most valuable guidance in these circumstances has to be strategic rather than tactical. It has to come from a broad understanding of what is happening. The actors in the drama need a 'mental model' of their world, located inside their own heads and conceived in their own terms. In the end, this will have to be based on deeper theoretical principles: for the moment, all I have to offer is this provisional analysis in the everyday language of organization folk.

1.4 How did it happen?

As we shall see, the causes of the transformation are quite complex and vary from country to country. For example, in the UK, until the end of the 1980s most active scientists would have described the various changes that had taken place over the previous 15 years as the unfortunate side effects of harsh economic policies, imposed by governments who were essentially obtuse to the long-term benefits of research. They saw present conditions as essentially 'abnormal', even though they had little evidence to suggest that there would be an eventual return to 'normal' conditions. In other words, it was all a question of resources: the troubles of science – like the troubles of the poor, as Bernard Shaw remarked – were due to not having enough money!

This interpretation is not unfounded, but it lacks historical depth. In 1963, Derek de Solla Price published a famous graph, showing that scientific activity had been expanding exponentially at a very high rate for about 300 years. He pointed out that if the publication of scientific papers went on growing like that, doubling every fifteen years, then soon every man, woman and child in the country would have to be spending all their time doing research and writing scientific papers. This was an absurdity: it was time to ask when the expansion would stop, and what would happen to science when it met its own limits to growth.

In simple terms, this is what we must have been seeing in recent years. Conventional econometric studies indicate that the R&D activity of a developed country now takes somewhere between 2 and 3 per cent of the national income. The precise figures, nation by nation, year upon year, sector by sector, are much in dispute, but with the possible exception of Japan they all began to stick at about this level in the mid 1970s and have risen very little since.

Some people argue that this was a pause associated with a world economic depression, and insist that scientific expansion has resumed, or will resume shortly, in some fortunate countries. The response to that must be that none of the optimists are seriously suggesting that scientific activity is set to double in the next fifteen years, let alone make up for the years when it stood still. Can they be expecting to reach a situation where, say, 10 per cent of the gross domestic product (GDP) goes on R&D, i.e. more than most countries are willing to spend on health, or defence? Tell that to the Financial Secretary to the Treasury, or his/her equivalent, in the next Public Expenditure Survey negotiations, or their equivalent. Explain that in your party manifesto, or even in your not-very-high-tech company prospectus!

There is no law of nature ruling out substantial further growth, but there are some pretty reliable principles of politics against it. The likely prospect is that science will have to exist for the future *within a fixed or slowly growing envelope of resources.* Some people are discouraged by this thought, or regard it as a damaging or disloyal admission of weakness. They point to this or that other country where a courageous, far-seeing government has raised the science budget by a few percentage points – and ignore a number of other countries where this item has (most reluctantly!) had to take the brunt of (absolutely unavoidable!) economies in public spending. In spite of considerable variations from place to place, and from time to time, there is good reason to refer to it broadly as a 'steady state' situation from a financial point of view, and to attribute many of the structural changes to this factor.

There are other factors, moreover, which are producing much the same effects. On the one hand, there are 'external' forces, exerted by society at large. Science has been so influential through its technological applications that it is expected to make greater contributions than ever before to the national wealth and welfare. This part of the story is so familiar that there is no need to go into it in detail. The concept of a national R&D system as a wealth-creating motor for the whole economy puts science under extreme pressure to be efficient and accountable. In other words, it must produce more value for the money it gets, especially in relation to presumed national needs.

On the other hand, all the old expansive forces are still at work within the scientific enterprise. As Derek de Solla Price pointed out, research has always been a growth industry. One can easily see why it has to double every fifteen years. The advance of knowledge opens up tempting opportunities for yet further advances. The solution to every scientific problem suggests two new questions that could now be tackled. A successful researcher never quite succeeds in working himself out of a job. Even when he retires, he wants two new jobs to be created, for his two most brilliant pupils to continue his mission. Old-established disciplines do not wither away: they split like amoebae into several new ones, or re-emerge triumphantly in novel interdisciplinary combinations.

The present era in science and technology also seems to be unusually fruitful. Perhaps every era has seemed the same to those living in and through it. Nevertheless, with the theoretical unification of the biological sciences around the mechanisms of heredity, and the instrumental sensitivity available through progress in information technology, the possibilities for both conceptual and practical advance seem limitless on almost every front. It is not just that many specialized fields of scientific effort may have reached the stage of 'finalization', i.e. the stage where there is a reliable theoretical framework to guide research strategically towards envisaged and desired ends. It is that startling progress is being made by the imaginative combination of concepts and techniques drawn from disciplines or sub-disciplines that were previously thought to be on distant continents of the academic map.

Needless to say, the interdisciplinary linkages and sophisticated instrumentation required to exploit these possibilities can be very expensive of money, time, and other resources. Quite apart from the financial and organizational lumpiness of indivisible Big Science facilities, R&D systems are under immense internal pressures to meet the increasing cost of the apparatus, technical services and administrative infrastructure required to do internationally competitive research.

These factors combine to generate a demand for resources for good-quality research – not only funds, but trained research personnel – that greatly exceeds the potential supply. From the standpoint of the scientific leader or research manager, it makes little difference whether the national aggregate of these resources has permanently levelled off in absolute terms, or whether it is growing at or slightly above the rate of growth of the national income. The 'steady state' condition may not be literally true but it correctly indicates a permanent climate of extreme resource scarcity for science.

1.5 The transition to 'steady state' conditions

Science in the UK and many other countries is thus under increasing stress from contrary forces. One can think of it as a highly expansive substance, contained in a cylinder with fixed walls, and pressed down heavily from the outside by a piston. The balance between these forces cannot be sustained for long without, so to speak, a chemical transformation in the substance under pressure.

Once again, a historical perspective is helpful. Modern science was 'invented' as a social institution in Western Europe in the seventeenth century. In the nineteenth century this institution was incorporated into 'academia': scientific research became a regular profession closely associated with other aspects of advanced scholarship and higher education. Despite variations from country to country in their official status, researchers in basic science worked everywhere under very similar conditions. They followed reputational careers in characteristically 'academic' organizations, held together largely by informal collegial relationships between autonomous individuals.

Although traditional academic institutional arrangements are the despair of modern management experts, they are not as fractious, fragmented or futile as their satirists depict them. They satisfy remarkably well the conflicting personal and corporate demands of a very subtle vocation. But these arrangements evolved by custom over a long period, *under conditions of continuous expansion*. They work on the tacit assumption that this expansion will never cease.

The intellectual dynamism of the whole scientific enterprise – the way that scientific knowledge and technological capabilities remake themselves every 25 years or so – has always been accommodated by institutional growth. It is taken for granted that tenure for the rising talent of today will not block the path of promotion for the next generation, that there will always be new posts for innovative individuals, that there will be the opportunities, the mobility and the resources to satisfy the ambitious intellectual entrepreneur, and so on. The long-established social structures and customary practices that channelled this dynamism worked very well for the best part of a century. The trouble is that they cannot bear the strains imposed by 'steady state' conditions.

In the UK, in particular, the transition to level funding was peculiarly disheartening and disruptive. For a time there was a serious crisis of morale, epitomised by the call to 'Save British Science'. This despondency about the future of science 'as we know it' was not widespread globally, but the scientific communities in many other countries have expressed similar concerns at the way that their governments have not only cut or limited research funding but also imposed new priorities and administrative complications on scientific activity. As Derek de Solla Price foresaw, academic science was so accustomed to being a growth industry that its whole ethos seemed to be at stake. Yet even he did not fully appreciate the power of all the forces pressing on the research system, and the extent of the structural changes that these were bound to produce.

1.6 Where are we heading?

The changes observed by our hypothetical space-time traveller are evidence of an irresistible historical process. Since the late 1960s we have witnessed a transition to a new regime for science, a regime that is better adapted to 'steady state' conditions. However much we may deplore many of its features, we are in a new ball game, and we must play it according to its own rules. These rules are themselves in a state of flux. To make sense of them, and shape our future within them, requires a cool, objective analysis, not a lament for a world of the past.

But there is a problem in presenting such an analysis. It is generally agreed that the historian of science can only understand past scientific episodes by trying to view them through the eyes of those who actually lived through them. Similarly, the present-day social analyst must try to view present-day institutions in the way that they are perceived by the majority of their members. This book attempts to describe the present regime in science by its own lights, in its own terms, according to its own rationale. In other words, I have tried to suppress the normative overtones that inevitably colour an account of the present by a person who can still recall different, older days.

Nevertheless, some of the things that are happening to science seem so out of keeping with its traditions that I do not find it easy to describe them dispassionately. They seem to sap at the very foundations of the whole enterprise. Could political demands for more and more in return for less and less prove counter-productive? Are the pressures on and within science likely, in the end, to reduce seriously the quality of what it produces? This must be of extreme concern to the whole scientific community. We cannot say for sure where science is going, but we can at least take every opportunity to push it in the right direction. As I have emphasized, that does not mean that a U-turn to something like the traditional system is called for, but it does require some appreciation of how that system actually worked.

Our only experience of a successful scientific culture is of the one that actually evolved, over a period of three centuries, in the heart of European civilization. What principles were *essential* to its achievements, and how might those principles be safeguarded as this culture itself undergoes radical change? That is why the last chapter of this book looks back, for a moment, at the intellectual attitudes and social practices that have sustained the credibility and reliability of scientific knowledge in the past. There was political design, not just sentiment, in Mark Antony's ironical claim that he 'came to bury Caesar, not to praise him'. Even as we bury many much-loved features of the scientific life, we must continue to praise and keep alive the 'method' that made them so fruitful.

A final word before plunging in. Notice that I have used the word 'science' without defining it, or naming its parts, or differentiating it from technology. This was not an oversight. As we shall see, one of the major characteristics of 'steady state' science is that all its outer and inner boundaries are open and indeterminate.

2

Scientific and technological progress

Aladdin rubbed the lamp. There was a great clap of thunder. A giant Genii appeared out of a cloud of smoke. 'Greetings, Master!' said the Genii, 'Thy every wish is my command.'

2.1 The unabated advance of knowledge

The origins of our science and technology lie in the mists of antiquity. Most of the essential features of the present way of doing science were already established in seventeenth- century Europe. People such as Isaac Newton and Anton van Leeuwenhoek devoted their lives to observation, experiment and theory, and contributed their discoveries to a common pool of knowledge. Institutions such as universities, academies, learned societies, and scientific journals provided the means by which scientists could publicly recognize or criticize each other's work. The knowledge acquired by research could thus be tested, consolidated and made the basis for further advances.

The systematic accumulation of knowledge by this process has been going on for more than three centuries. We look back now in astonishment at how far we have come. How exciting it must have been to hear for the first time about the spiral galaxies, the age of the dinosaurs, the germs that carry disease, or the atomic nucleus. In all that time, there has been no significant break in the history of science as a human undertaking. Scientists still celebrate the lives of their great predecessors almost as if they were former colleagues – earlier members of the firm, now regrettably retired and, alas, passed on.

All our modern technologies – engineering, agriculture, medicine, political and commercial practices, etc. – have similar long histories of steady technical advance. Sometimes this advance has been closely linked with new scientific discoveries: at other times, quite novel techniques have been invented with little reference to contemporary scientific ideas. Again, looking back, we can only marvel at what has been achieved, in just a few centuries, for the health, wealth and material well-being of mankind. Yet we also recognize that this is merely the outcome of a continuous process of mobilizing and combining a multitude of skills, concepts and products in the pursuit of technological progress.

The advance of knowledge and the improvement of useful techniques continue world-wide on an unprecedented scale. It is very difficult, of course, to judge whether what is now being discovered or invented is as novel and significant as what was discovered or invented in the past. By what criteria ought one to compare James Watt's first steam engine with Frank Whittle's first turbojet? Are the basic principles of plate tectonics, which were only worked out in the middle of the twentieth century, any less fundamental than the basic principles of stratigraphy that enabled nineteenth century geologists to infer the long history of the earth?

There seems to be a healthy tendency to undervalue present-day achievements. The natural desire to celebrate a recent discovery that seems at last to have solved an existing problem is inevitably outweighed by failure to appreciate other work with crucial implications for problems that have not yet been formulated. In science and technology, as in the arts and literature – indeed, as on the stage of history itself – posterity is the final arbiter of excellence.

The most we can say is that there is no serious evidence that science has changed or is about to change its nature. For good or ill, it remains, as it has always been, an essentially open-ended and progressive enterprise. That is to say, there are no signs of limitation or saturation in the extension of what is now known, or could be studied. Astronomy, for example, is one of the oldest of the natural sciences, yet there has been no period in the past to parallel the last twenty years for the quite unsuspected discoveries that have been made about the structure of the Universe.

The same applies, in its own sphere, to every branch of technology: we cannot do everything that we would like or could imagine, but there is not a closed boundary around what can now be achieved technically, or might be achieved technically, given the means. To take an obvious example, medicine will surely see just as much progress in the next 50 years as it has since the 1930s: we can only guess at what the most important new developments will be, but they can scarcely prove less significant than, say, the introduction of antibiotics, and routine organ transplantation, in the previous period.

The changes that are now taking place in the way that scientific activity is organized cannot be attributed to the fact that it is approaching some natural limit to its growth. Science as a whole is not bounded in the way that, say, the total population of the earth must be bounded by the land space on which people can dwell and grow their food. We may have our doubts about the need or desirability of pursuing certain lines of enquiry. We may decide, in our wisdom, not to develop some particular technology. We – that is, the human species – might even, through some supreme folly or mischance, destroy our civilisation, forget what we have learnt over the centuries, and revert culturally to a Darker Age. But in science itself we are certainly not yet in the position of Alexander the Great, lamenting that he had no more worlds to conquer.

2.2 Scientific change

The Siamese twin processes of scientific discovery and technological invention go on as in the past. As knowledge advances, however, it does not merely expand in quantity: it changes its nature. Research as a 'method' stays the same, but the frontiers of knowledge and technique are always on the move into new territory. The research problems of today are not only very different in substance from what they were only half a century ago: they are also different in their intellectual character, and in the methods used to tackle them.

Most people are aware that certain fields of knowledge have been 'revolutionized' during the last half century. Take, for example, geology. From the early 1800s to the 1950s, it seemed a useful but rather staid science. Geologists just went on quietly building up an enormous mass of information about the current composition and distribution of rocks and fossils. This information was systematically interpreted on the very reasonable assumption that the general map of the earth had not altered substantially in five billion years. Then it was realized that the continents themselves had broken up, moved about on the globe and recombined a number of times – and that this process was still going on. Almost all that was known about the history of the earth had to be reconsidered. The theory of plate tectonics not only explained earthquakes: it was itself an intellectual earthquake. The old family mansion of the earth sciences was shaken down, and has had to be rebuilt to an entirely new, more progressive design.

It is not always appreciated, however, that almost every field of science has undergone similar changes, during the same period, if not quite on the same grand scale. Many subfields of knowledge have been deconstructed and reconstructed to take account of quite unforeseen discoveries or insights. In other cases, a few decades of apparently modest progress can add up to a complete transformation. For example, a series of patient, unspectacular observations of animals in the wild have profoundly altered our understanding of their social interactions, and hence, by implication, our interpretation of many aspects of human behaviour. Some subjects, especially in the social sciences, seem so perennial that one tends to think of their history as little more than a cycle of intellectual fashions: yet closer scrutiny would show that one would seldom really 'come out by that same door as in I went'.

Radical cognitive change is nothing new in the history of science. Go back from the 1930s another half century, to the 1880s; the sense of having entered yet another, fundamentally different intellectual world will be just as striking. Then read your way carefully into the knowledge and ignorance, the thought processes and unvoiced assumptions, of the early nineteenth century: it will seem quite different again.

What happens is that major discoveries, by opening up quite new fields of study, or offering quite novel technological opportunities, generate many more researchable problems than the enigmas they resolve. As existing fields of research seem to become less fruitful, their place at the advancing edge is taken by more fertile new ones. These expand vigorously, their insights and techniques diffusing widely across the scientific map until they seem to dominate the scientific thought of a whole period. This happened in the middle of the nineteenth century, with the ideas of Energy and of Evolution. We can see the same thing happening now, as our recently acquired understanding of the structure and function of DNA has transformed basic and applied research in all the biomedical sciences and technologies.

The question is whether some of these conceptual changes are so significant that they represent a radically new approach to scientific questions. The 'scientific method' itself must continually evolve to adapt to the new intellectual and social environment that it is helping to create. The central theme of this book is the profound transformation that is taking place in the *institutional* arrangements for undertaking research and invention. To what extent is this transformation due to a general change in the *type* of knowledge now being sought, quite apart from changes in the means for seeking this knowledge?

Such questions about the spirit of the times are notoriously difficult to define, let alone to answer meaningfully. Nevertheless, it does seem safe to say that the scientific world map does not have so many open areas of almost complete ignorance as it used to. There are, of course, many small regions that are totally blank. There are also immense continents still to be explored: the development and working of the brain, the origins of life on earth, the unification of the forces of physics, etc., not to mention behavioral and social enigmas such as the links between crime and punishment or the economics of boom and bust. But the smaller regions of ignorance are bordered by highways of understanding, and we think we have located most of the unknown continents, have roughly charted their coasts, and have already dispatched a number of expeditions into their interiors.

Scientific knowledge is not only growing rapidly in its scope: it is also becoming more tightly interconnected. Theoretical or practical problems that were once studied independently within distinct disciplines can be seen to be related. Surprising linkages can thus develop between very distant bodies of knowledge or technique. Computer programs developed for X-ray scanning of the brain turn out to be applicable to seismic prospecting for oilfields. Biological enzymes are found to be ideal as catalysts in heavy chemical plants. The microscopic study of radiation damage in plastic films leads to an effective means of detecting cosmic rays. The list of unforeseen connections is endless.

The linking of apparently unrelated pieces of information, technique or theory has always been one of the most creative processes within science. In the past, however, it was often limited by the lack of coherence and continuity between the intellectual schemes developed in various academic disciplines. The goal of unifying all the sciences was seldom more than a slogan. A stroke of imaginative genius was required, for example, to link the economic theories of Thomas Malthus with the problem of biological evolution, or to see the significance of non-Euclidean geometry for an understanding of gravitation.

The sciences are still far from unified theoretically, but their complex, multiple interconnectivity manifests itself in the familiar phenomenon of 'spin-off'. This is a much more general phenomenon than would be suggested by such stereotypes as the development of the non-stick frying pan from the material used for the re-entry nose cone of a moon rocket, or the evolution of the pocket calculator from the compact electronic circuitry required for weapon systems. Spin-off is seldom a one-way process. Weapons systems nowadays use the micro-chips developed commercially for home computers: the Apollo programme was surely indebted to the know-how of the kitchen utensil industry. Knowledge and technology transfer can now proceed in any direction – from applied science to basic, from civil to military, from the microscopic to the telescopic, from psychology to pure mathematics and back, across the whole spectrum of the sciences.

Another consequence of increased connectivity is that problems are often approachable from many different directions. Scientific knowledge now tends to grow particularly vigorously in *interdisciplinary* areas, or to make particularly striking progress when it can be fitted together into a coherent, *multidisciplinary*, conceptual scheme. Thus, for example, the relatively narrow and specialized subject of linguistics is thriving nowadays on a mixture of fundamental concepts imported from the traditional humanities, the social sciences, the biomedical sciences, and the new world of information technology

2.3 The increasing applicability of 'finalized' science

The development of a consistent network of reliable concepts covering a particular domain of science – what Thomas Kuhn taught us to call a 'paradigm' – induces a decisive change in the nature of research in that domain. It makes it possible to envisage a variety of further advances, either of knowledge or of practical application, before they have actually occurred. We can see this very clearly in the case of molecular biology. The elucidation of the function and structure of DNA not only opened the way to many important discoveries in biology: it encouraged researchers and other biomedical workers to think about the role of DNA in diseases such as cancer, to conceive new ways in which such diseases might be attacked, and to design research programmes to test and develop these ideas. In other words, the *science* of molecular biology is becoming the basis for bio-*technology* or genetic *engineering*.

One way of reading the history of science has every discipline passing through a series of stages, from the 'pre-paradigmatic' stage, where speculation is rife, and almost anything goes, to a state where research can be 'finalized' – that is, where research programmes can be set up with quite specific theoretical or practical goals. In classical fluid dynamics, for example, a stage has been reached where the basic equations of motion are no longer controversial, and where there are well-established techniques for computing the outcome of almost any situation and testing such predictions experimentally. This does not mean that the subject has been 'reduced to first principles' and thus exhausted of scientific interest. Quite apart from endless possibilities for effective application, there is still a great deal to be learnt by imagining novel physical situations, and interpreting the phenomena that can arise in them, theoretically or experimentally.

It must be admitted, however, that this reading of the history of science is much disputed. Many fields of science may reasonably be described as having been finalized in recent years, but that does not mean that the paradigms that govern them are quite secure, nor that every scientific discipline is fated eventually to reach this last triumphal stage. In practice, as the recent discovery of near room-temperature superconductors dramatically illustrates, further exploration sometimes leads to completely unexpected discoveries or suggests many quite unforeseen applications. The notion of having reached a state of finalization may be a complacent illusion. In the late nineteenth century many physicists already believed that their science was complete, and that all they had left to do was to add another decimal place to their measurements: they had no idea that there were all the properties of atoms, and nuclei, of nucleons and quarks, still waiting to be uncovered. And how does this notion apply to the social sciences, where the practical implementation of theoretical understanding of a social situation is very much more complex and subtle than it usually is in the natural sciences?

Nevertheless, it is characteristic of many branches of science today that they count themselves as effectively in this finalized state. This has a very important influence on the way that research is conducted. In a supposedly finalized science, such as fluid dynamics, or plant genetics – or even economics – it becomes almost impossible to make a sharp distinction between 'basic' and 'applied' research in terms of what is actually discovered. For example, it might be very difficult to determine from a highly mathematical analysis of turbulence whether it had originally been motivated by intellectual curiosity, or whether it had been undertaken in order to solve an urgent practical problem arising in the design of an aircraft. Similarly, experimental phenomena reported during a process of technological development – for example, in the long-sustained effort to breed a commercial variety of hybrid corn – might equally well have been observed in the course of research designed solely to improve scientific understanding of the underlying mechanisms.

This is not to say that all research is then undertaken for the same purpose. The personal motives and organizational goals behind particular research projects may still be very diverse. The professor of applied mathematics chooses to work on the theory of turbulence in the hope of eventually being able to publish some interesting results that will add to his scientific reputation: the aeronautical engineer is instructed by the management of his firm to work on the same theory in the hope that it will lead them to a more efficient design. But this diversity will not be obvious in the form of the question being posed, the means being used to tackle it, or the nature of the answer.

To appreciate this point, imagine being transported blindfold into an unknown building, and then observing the work going on around you. People in white coats or roll-neck sweaters are seen looking down microscopes, running ultracentrifuges and gas chromatographs, putting racks of test-tubes into incubators, keyboarding data into computers, consulting articles in journals whose titles include words such as 'cell' and 'gene', and arguing incessantly about 'reverse transcriptase', 'introns', 'homologous sequences', etc. Clearly, you are in a laboratory devoted to research in molecular biology; but is it in a university or in a commercial firm? Weeks might go by before a clue is given: somebody says 'We won't get a Nobel Prize for this discovery if we don't publish it promptly' or, alternatively 'This will give us the product we need to sustain our stock-market rating.'

As we shall see, the convergence of basic and applied research at the level of the laboratory bench has a profound effect on the way that science is organized. In many fields, for example, the time required to exploit a fundamental discovery industrially has become so short that commercial firms cannot afford to wait until the results of academic research have been published before setting about trying to apply them. They have to be directly involved in basic research themselves, or keep a very close watch on what is going on in universities all over the world, in case something interesting turns up. At the same time, academic scientists are increasingly aware that a new industry might possibly arise out of their work, and are advised to take out patents, or to set up small businesses to market their products.

In very broad terms, the effect of finalization is to dissolve the technical interfaces between a scientific discipline and the technology (or technologies) where it is put to practical use. This does not mean that 'science' and 'technology' are becoming identical in other respects. As scientists and technologists themselves insist, they differ profoundly in their goals and in their social roles. For example, the job of the micro-electronic engineer is not the same as the job of the solid state physicist. It requires different training and is directed to different ends. Nevertheless, it is significant for policy that these two professions largely share the same basic concepts, draw upon the same data, use the same instruments, come up against the same technical obstacles, and are often employed side by side in the same R&D teams.

2.4 Expanding opportunities

The large areas of science that are now being staked out for finalization present new technological opportunities on an unprecedented scale. An outstanding example is solid state physics. In the 1930s, quantum theory began to make sense of the characteristic properties of metals and insulators – why copper was a good conductor of electricity and diamond was not. This theory also explained, in principle, the intermediate case of 'semiconductors', such as silicon, whose peculiar properties were explored experimentally. In 1948, these were made the basis of a novel electronic device: the junction transistor. In the next 40 years, this device went through a whole sequence of technological developments. It was speeded up, miniaturized, radically modified in basic structure, integrated into arrays, and mass produced in vast numbers. In the form of the ubiquitous 'microchip', it has become the mainspring of all our information and control technologies.

But solid state physics did not stand still during that period. The complex electronic properties of particular metals and semiconductors have been explored in detail, and explained quantitatively in terms of their atomic constituents. One can now understand, for example, why it is that the electrons in a compound of indium and antimony are much more mobile than they are in silicon, making it a better material for certain devices. Again, the behaviour of the electrons in the vicinity of various types of impurity or crystal imperfection has been studied, so that one can work out the effects of 'doping' a particular semiconductor with minute quantities of another element. A whole variety of other physical phenomena – optical properties, thermal properties, high field phenomena, low temperature behaviour, magnetic properties, surface properties, the behaviour of material in very thin films, etc. – have been observed, measured, probed experimentally and explained theoretically.

This enormous advance in basic understanding of the physics of solids has not only benefited the development of the transistor: it has opened the windows of imagination on to a vast landscape of quite new devices, working on very different principles and adaptable to very different purposes. Indeed, many of the components of the modern microchip are not really junction transistors at all, but depend for their operation on some other phenomenon, such as a 'field effect', or 'charge coupling', or 'impact ionization'. Solid state devices based upon superconductivity, laser action, 'low-dimensional quantum effects', etc. are now envisaged, using very much less power, working at much higher speeds, and able to perform some of the very complex operations required for signal processing, pattern recognition, data storage, and so on. In combination with parallel developments in optics, solid state physics is also behind the proliferation of technology into opto-electronics, which is already revolutionizing the telecommunications industry.

The current cycle of high-technology industry is centred on information technology, growing out of basic discoveries in electronics, solid state physics, materials science and mathematics. This cycle is still, apparently, in a phase of rapid expansion, with no signs of approaching saturation in invention or investment. Science and industry are already preparing for similarly rapid growth over a wide area centred on 'biotechnology', born of fundamental developments in genetics, biochemistry and microbiology. Beyond that, it is thought that a further cycle may already be germinating out of basic research in psychology, physiology, information technology and molecular biology. In all of these disciplines there have been impressive cognitive developments opening the way to finalization and practical exploitation.

This celebration of progress on the scientific and technological frontiers is not inserted here simply out of euphoria. It is to remind us of the immense range of realistic opportunities for further technological development thus presented. The flying machines and submarines in Leonardo da Vinci's notebooks were mere fantasies, quite incapable of realization in anything like the forms he suggested: few of the innumerable devices now being sketched out by the tough-minded engineers in the electronics industry may eventually appear in marketable form, but almost any one of them could be made to work in some sort of fashion if enough resources were thrown into its development.

The trouble is that these resources may have to be very large indeed to ensure success. An immense amount of further research and development may be required to exploit the practical potential of a newly discovered area of scientific knowledge. This activity may not only cost a great deal of money: it may also involve unusual managerial skill to mobilize and orchestrate the range of expertise needed to turn a concept into a reliable product. The development process is made feasible by the co-existence of this expertise in many different specialties – in the case of a new microchip, think not only of solid state physics and electronic engineering but also of all the chemical knowledge and experience that goes into material purification, crystal growth, surface preparation, masking and etching of circuit patterns, implantation or diffusion of dopants, attaching contacts to the chip and canning it safely.

The advance of knowledge in many fields of science and technology has thus put us into a position where the choice of the next step forward is less constrained by proven technical means than by the desirability of the ends that might then be achieved. We are a bit like railway engineers at the end of the nineteenth century. It was technically possible – at a cost – to build a line across almost any terrain, however mountainous or swampy. The real question was whether the very large and indivisible investment required to undertake and complete any particular project would eventually yield a profit.

As the above simile suggests, this is not an entirely novel situation. But some of the projects now being seriously undertaken or entertained by governments or big firms are not just very large and technically audacious: they are unprecedented for their confidence that ways will be found to get around difficulties that seem impassable by present methods. The significant parameter is not the total investment required: the building of a pyramid must have beggared Egypt for a generation, and the effective cost of one of the great European cathedrals must be reckoned as comparable with the construction of the Channel Tunnel. Nor is it sheer boldness in extrapolating from existing designs: engineers such as Brunel undertook vast projects that went far beyond the accepted principles of their contemporaries. The novel element is the projection of the enterprise beyond the state of the art into a whole complex of techniques which do not yet exist, but which will presumably be rendered feasible by discoveries that are conceivable but have not even been made.

There is now extraordinary confidence in the sheer power of science to clear a path towards any desired objective, once this seems attainable in principle. This confidence springs, perhaps, from the outstanding technical success of the Manhattan Project, which took little more than five years to transform the back-of-envelope ideas of a few academic scientists into an overpowering weapon. It is continually exemplified in the development of other high-technology systems, under the sea, in the air, and in outer space. As the Strategic Defence Initiative of the United States Government clearly demonstrates, apparently responsible people can now be persuaded that targeted R&D can blast its way to any technological goal, however implausible this may seem to the great majority of the relevant experts.

It is easy, of course, to offer objections to uncritical technological optimism, even when its goals are perfectly benign. An obvious counter-example is the story of President Nixon's 'War on Cancer', in which hundreds of millions of dollars of biomedical research failed to make a breakthrough to a cure. The horizon of the world-wide programme for obtaining electric power from controlled nuclear fusion recedes steadily, year for year, into the future, despite the sustained efforts of numerous extremely able scientists and engineers over a period of more than three decades. Many apparently straightforward technological developments – the fuel cell, machine translation of natural language, a vaccine against malaria, etc. – have been held up by unforeseen difficulties that no amount of research seems able to resolve.

But these cases of technical failure simply demonstrate statistically that technological innovation is still a very risky business. The fact that some beautiful ideas just don't work in practice, however much is spent in trying to get the science right, does not deter people from believing in the instrumental power of R&D. This belief is held by the opponent of economic growth as well as by the venture capitalist, by the disarmer as well as by the weapons buff, by the advocate of research on health and social welfare as well as by the enthusiast for wealth creation. There is general agreement that science puts into the hands of powerful institutions a highly effective means of moving towards quite distant practical ends, however desirable or undesirable these ends may be.

The advance of knowledge as such thus explains one of the major features of the present situation: the emphasis on 'market pull' as the primary motivating force in the planning of science. This is not simply a reflection of current socio-political partiality for commerce and industry. It applies to the whole tendency to assign priorities to research programmes in proportion to the supposed practical value of what they might produce, for health and happiness as well as for work and wealth.

The instrumental conception of science goes back at least as far as Francis Bacon. But its current dominance in every aspect of science policy is not just a contemporary fad. We really seem to have reached a position where we can do far more than we can afford with the knowledge we have, or could probably get if we sought it. We therefore feel that we should put our money where we think we will really get the best out of it, rather than venturing it on a variety of enterprises of which few are likely to prove fruitful.

2.5 The rise of the concept of 'strategic' research

The defect of 'market pull' as a paramount principle of science policy is that it underestimates the inherent unpredictability of scientific discovery. It takes at face value the speculative mappings of regions of knowledge that have never been carefully surveyed. It fails to allow for the unforeseeable – almost unimaginable – practical capabilities that would also continue to flow from science even if it were motivated mainly by curiosity or technical virtuosity.

As an insurance against this risk, increasing stress is now laid on the concept of 'strategic' research. The idea is that the tactics of applying science for immediate, short-term purposes need to be broadened into a more general strategy of undertaking more basic research with wider, longer-term goals. In other words, the long-standing distinction between 'pure', or 'basic' science, on the one hand, and 'applied' science, on the other, is losing its meaning. Even in a finalized domain of science, it clearly makes good sense to foster open-ended research that is not directed towards achieving particular practical objectives, but which is likely to strengthen the supporting framework of understanding, or extend the linkages to other branches of knowledge.

An obvious example of a strategic research area is the new science of materials – not just metals and insulators but ceramics, high-density polymers, metallic glasses, fibre-reinforced composites, amorphous semiconductors, bio-compatible alloys and many, many others. In the past, most of the research that was done on the structural properties of materials such as engineering alloys was essentially empirical. There was very little general theory to guide the necessary work of amassing data, formulating rule-of-thumb recipes, and developing reliable testing procedures for the new materials required for particular uses, such as motor car bodies or gas turbine blades. Until about 50 years ago, many of the simplest properties – for example, the brittleness of cast iron – were apparently inexplicable. But the advent of the theory of crystal dislocations in the 1930s made it possible (at least in principle) to predict the properties of new materials from knowledge of their chemical composition and crystal structure – or, rather, to explain the observed properties in detail and suggest practical ways of modifying them as desired.

A new scientific discipline has thus come into being, unified by this theory and its subsequent elaborations. Powerful experimental techniques such as electron microscopy have extended the scope of this theory to the whole field of structural materials. It is scarcely necessary to emphasize the value of the technological applications of this discipline to the improvement of existing materials, or the development of relatively new ones. But a vast range of more academic questions has also been opened up for research: how, precisely, are the atoms arranged in a metallic glass; what makes the chain molecules in a crystalline polymer fold up the way they do; can the mechanical properties of composite materials be described by a simple mathematical formula; what happens at the boundary between a piece of metal and a living tissue – and so on, almost without end.

In one sense, this research, which is not directed towards the solution of any specific technological problem, could be described as 'basic', in that it is usually related to answering general questions about characteristic phenomena or underlying mechanisms. It is also 'academic', in that the immediate goals of research are set by the researchers themselves, motivated (as they would say) by 'curiosity', or (as others might suggest) by the professional necessity of making a publishable contribution to the subject. Many departments of materials science in universities and other quasi-academic institutions are, in fact, genuinely engaged in trying to extend, deepen and elaborate our understanding of such matters, without any reference to their practical use.

Experience has shown, however, that this research is very far from useless. In the long run, it makes a great contribution to the more direct applications of this branch of science. One can see, for example, that a proper understanding of the geometry of polymer crystallization, or of the behaviour of composite materials under stress, or of the interaction between a living cell and a metallic surface, could be a vital piece of information in the search for a new type of material for a quiet-running gear wheel, or an aircraft wing, or an artificial hip joint. Even though specific applications are not foreseen when the research is begun, many potential applications of the results can easily be envisaged by those who undertake or sponsor it.

Of course, as always in research, a substantial proportion of such projects will eventually turn out to be fruitless. They may fail to prove anything scientifically, or they may produce results that cannot be applied technologically, or they may only suggest inventions that cannot be marketed. Naturally, one would like to be able to assess the ultimate profitability of each research proposal before it is begun. But this would only come at the end of a long trail, stretching perhaps twenty years ahead and interwoven with the results of many other unforeseen discoveries, which cannot be plotted in advance. At this stage, the precise degree of apparent technological relevance of the proposed research is not of great significance. The important issues are whether the initial question is well posed, the likely results scientifically interesting, the research method appropriate, and the researcher really competent to carry out what is intended.

Strategic research is often very closely connected with the broad development of a 'generic' or 'enabling' technology. For example, research on a basic mechanism of semiconductivity may merge imperceptibly into the detailed investigation of the behaviour of carefully contrived configurations of impurities, etc. on a silicon chip – configurations which are being studied simply because they look as if they might soon make possible a whole new range of micro-electronic devices. Similarly, strategic research in molecular biology underpins the very general technology of gene transfer, which 'enables' many quite different processes, such as manufacturing insulin from genetically engineered micro-organisms, or breeding a frost-resistant variety of tomato. The scope of strategic research can thus stretch a long way, into the broad fields of what used to be called, loosely, 'applied' science.

The concept of strategic research can also be extended a long way in the other direction towards the 'pure' end of the R&D spectrum. Interpreted literally, it provides a rationale for supporting a large part of all academic research in the natural and social sciences. There are few subjects that cannot be justified in principle for their 'strategic' relationship to progress in one or another of the industrial, medical or agricultural technologies, not to mention the arts of management and government. It is hard nowadays to imagine any approach to biology, however 'fundamental', that might not yield medical benefits. A subject such as palaeontology, dealing apparently with a remote period of the past, has its implications for petroleum recovery. The theory of numbers, supposedly the least applicable branch of pure mathematics, has made a highly effective contribution recently to the ancient craft of cryptography. Can there be any hope of coping with crime without a much deeper understanding of the wellsprings of human behaviour?

In practice, the use of this terminology revolves around the plausibility of the supposed potential for ultimate application [§8.3]. In the current parlance of science policy the term 'strategic' is supposed to be restricted to 'generic research on potentially rewarding disciplines or technologies', indicating that a managerial rein is being kept on sheer curiosity. But recent history strongly suggests that the definition of this potential can be very elastic [§5.7]. The basic research laboratory is surely still the place where the next innovatory cycle of high technology will be conceived. But there remains a wide gap between the boundless imagination of those with an interest in doing the research and the more sober judgement of those who are supporting the research because they think they might profit from its results!

2.6 The role of 'pure' science

If we are thinking of what might happen scientifically and technologically out to the most distant horizon – 50 years, say – then the category of strategic research is very broad indeed. Nevertheless, it cannot reasonably be stretched to cover all the science now supported on a large scale from public funds. Instrumentally sophisticated subjects, such as high-energy particle physics, astrophysics, cosmology, or planetary science can earn a large part of their keep through their 'spin off' in advanced techniques and skilled personnel. Humanistic disciplines such as archaeology and ancient history are not very expensive, and play a vital part in defining the frame within which we now live. But any argument for supporting them for their potential contributions to 'wealth creation' is bound to ring somewhat hollow.

It is easy to say, of course, that one really cannot write off completely all the applications that might eventually flow out of an area of knowledge that has not yet been explored. This point has already been emphasized several times, and is fully supported by the experience of history. Nevertheless, when it is necessary to make hard choices involving the expenditure of millions of pounds – as, for example, in deciding what proportion of the UK science budget should go to high-energy particle physics at CERN, the great European accelerator complex near Geneva – utilitarian considerations have to be taken seriously, and cannot be simply brushed aside with the rhetorical question: 'What is the use of a new born baby?' It must be accepted that there are fields of science for which any eventual practical applications are scarcely conceivable, and which even the most liberal accountancy would discount to zero present value in this particular respect.

A shadowy line of demarcation is thus developing across what used to be called basic or fundamental research, leaving a substantial part of academic science outside the shelter of the strategic umbrella. Some of this research, for example in astrophysics, is so vivid and dramatic that it engages widespread public attention and support. The strength of the 'Gee Whizz factor' is not to be denied. Some, as in high-energy particle physics, is extremely recondite, but produces significant steps forward in the long march towards an understanding of the fundamental forces of nature. It must be admitted, however, that the intrinsic attraction of much of this sort of research is only apparent to the scientists who specialize in it – and who have an obvious interest in continuing to pursue it.

Nobody doubts that research of this kind ought to and will continue – but on what basis and on what scale? The rationale of what might fairly be called 'pure' science, has now become a very serious issue for the scientific world. It cannot be answered satisfactorily without an appeal to deeper social values than conventional notions of practical utility. But to say that pure science and other forms of scholarship should be supported simply 'for cultural reasons' merely begs the question. To insist that 'human curiosity must be gratified' is descriptive rather than evaluative. Without the argument for potential utility at its centre, the whole case tends to become very soft and open.

But there is another way of looking at this issue which the hard-headed realist should consider very carefully. It involves some appreciation of the nature of the scientific enterprise, and the tradition that it has developed over the centuries. Within that tradition, it is not hollow rhetoric to say that scientists are personally motivated by curiosity and the search for the truth. Whether or not they are born with these traits, they are socialized into a community publicly dedicated to these aims and organized to further them. Having been trained to put the acquisition of scientific understanding above immediate personal or social expediency, they are entrusted individually with the task of advancing knowledge in particular directions on behalf of society as a whole – and rewarded for success in doing so.

As we shall see, contemporary changes in the way that research is organized are putting this tradition under pressure at a number of points. And yet the whole concept of strategic research relies upon it heavily. Even in a field such as molecular biology, with immense possibilities for application, it is essential that there should be many researchers carrying on their work as if motivated solely by curiosity. They will be expected to become personally committed to finding the answers to certain extremely specialized questions of no obvious utility, and to strive with one another in getting those answers right. They will be expected, moreover, to develop very high levels of expertise in judging what questions are worth asking from a strictly *scientific* point of view, and whether they are answerable by research. In other words, many of the most valuable results of strategic research will actually be produced by academic scientists who go on developing their subject with the same intensity, and by the same criteria, as they would if the advancement of basic knowledge and understanding were their sole purpose.

The line of demarcation between basic strategic research and pure research is inferred from a view towards a distant, misty horizon from a high point of vantage. It does not show up on the ground. A distinction in principle between potentially useful and essentially useless knowledge is of little significance at the laboratory bench or in the scientific treatise. The sheer effort of trying to draw such a distinction would itself constitute a major piece of research – not necessarily futile, but inevitably inconclusive. At what distance from the earth, for example, does basic space research lose its strategic value and pass over into pure astrophysics? At what level of abstraction does the attempt to formalize quantum theory, and draw philosophical conclusions from it, lose all contact with reality? At what epoch does ancient history become modern, and begin to influence our thinking about our social institutions?

These are not just quibbles, designed to support the notion that scientists and other scholars should be quite free to do whatever research they fancy, at the taxpayers' expense. They do not negate the necessity of giving a great deal of weight to all the practical potentialities and opportunities that research may offer. But they remind us, once again, of the unity, the inseparability, the interpenetration, of all branches of learning. The various disciplines and sub-disciplines of the sciences and the humanities advance hand in hand, if not quite in step with one another. The suppression of further progress in an active field of pure research would eventually have serious repercussions in the other fields to which it is linked intellectually, technically, or by professional tradition.

The role of pure science nowadays is thus much more complex and subtle than can be settled by a straightforward confrontation between utilitarian and cultural values. It involves questions about the criteria that scientists employ in assessing genuine progress in a particular subject, and about the level of support needed to keep that subject intellectually buoyant and healthy. It involves the motives that draw able young people strongly into such fields, and the wider applicability of the research training they may thus get. This is really a crucial issue, to which we shall be returning frequently throughout this work.

2.7 Scientific information as a commodity

A lot of ink has been spilt by philosophers, historians, and sociologists of science and technology in their disputes over where to start in their account of the processes of discovery and invention. Should they be approached in terms of the knowledge and know-how they generate – that is, in terms of theory and experiment, conjecture and refutation, concept, design, trial and demonstration? Or is the real 'scientific method' akin to 'inventiveness', or 'artistic creativity', and only to be understood through a deep study of the workings of the individual human mind? Or will we find that we can make little sense of science and technology as human activities without detailed reference to other aspects of the surrounding culture – its religion, its politics, its economics?

What can be said without dispute is that science is a *social* activity. Research itself may, on occasion, be a very lonely, highly personal occupation. But the results of research do not become scientific information until they have been communicated to other researchers. Open channels of communication within an attentive community are an essential condition for active scientific discovery and technological invention.

European science has grown up around the academic system of scholarly communication, where scientists are induced to publish their results as promptly as possible by the prospect of receiving personal 'recognition' and other rewards, tangible or intangible, for their discoveries. The flow of information through the traditional media of scientific journals and books is driven by competition for priority, and monitored by 'peer review' – that is, by critical appraisal, before publication, by other specialists in the same field. One might say, indeed, that these procedures are at the very core of academic science as a social institution, and are inseparable from its other functions [§7.5].

The transmission of information between technologists is not organized on such a systematic basis. Indeed, the presumption has always been that the knowledge embodied in a useful invention is essentially the private property of the inventor, to be held secret for further exploitation or sale. Nevertheless, technological progress depends on making this knowledge more widely available. One of the major functions of an official patent is to induce the inventor to disclose valuable information in return for a temporary monopoly on its use for profit. Legislation in favour of industrial and commercial competition has a similar effect of encouraging the rigorous testing of technological innovations through success or failure in the marketplace, and their rapid diffusion by purchase, imitation or inventive simulation.

The actual boundary between science as *public* knowledge and technology as *proprietary* knowledge has never been sharply defined. Scientific information with obvious technical implications – for example, the chemical composition and process of manufacture of a new dyestuff or explosive – has always been treated as a marketable commodity. But, the extension of higher education into advanced technologies such as engineering and medicine required that many trade secrets be shared and made public. At the same time, the potential utility of most academic research was usually so uncertain that a convention of complete openness became the norm. Any research result obtained in a university laboratory or similar academic institution was thus to be treated as a 'contribution to human knowledge' and published freely, even though it might contain the germs of commercially exploitable invention. One could say, in fact, that the term 'academic' traditionally signified that this convention was being followed.

The difficulty of maintaining this convention is one of the signs of a major change in the internal structure of science. In a finalized domain of science, information about the results of quite basic research can quickly acquire commercial or military value. An unexpected discovery – for example, a successful recipe for making a room-temperature superconductor – could lead within a few years to important industrial applications. Although it is seldom possible to keep the basic principles of such a discovery completely secret until they have been embodied in a useful product, a lead time of a few months may be decisive in a highly competitive market. The discoverer thus has strong incentives to keep this information quiet until it can be exploited profitably, or to sell the right to use it for a good price.

Every field of strategic research is subject to the same influences. If the research is funded because of its potential applicability, then there is no denying that its results might be of commercial value. Even when it is carried out in an academic institution, and motivated primarily by curiosity, research is now seen as a process of generating 'intellectual property', whose ownership needs to be established and protected in advance.

We find, for example, that an industrial firm or government agency sponsoring basic research in a university department may insist on having a first sight of all reports of the research, with the option of delaying the publication of some of them for a certain period. What then about research that unexpectedly turns up results of potential *military* value, which ought perhaps to be kept secret indefinitely in the interest of national security? As the discovery of the applicability of pure mathematical number theory in military and commercial cryptography has shown, this is not a fanciful case.

Many universities are laying down rules to cover such situations, often asserting their right as an institution to share in the patent royalties earned by their academic staffs. Changes in the law are taking place, to deal with the financial and legal considerations that now arise in deciding between individual and institutional rights to this property and how it should be disposed of. The growing recognition of the potential exploitability of even the most basic discoveries is thus transforming the rules and practices governing access to all scientific and technological information.

This change in the socio-legal status of basic scientific knowledge has far-reaching implications. The free flow of scientific information has always been the norm in academic science. This norm is not always obeyed in practice: as in the well-known story of the discovery of the structure of DNA, scientific groups often clam up completely when they are competing fiercely to be first with an important discovery. But this secretiveness is strictly limited, since they must publish their work promptly if they are to gain the credit for it. A delicate balance has thus developed, with all the ethical weight on the side of openness. A shift in this balance, towards secrecy, not only slows the advance of knowledge: it also puts a damper on public assessments of research claims, which are the ultimate arbiter of scientific validity.

3
Sophistication and collectivization

When the artificial bird had been wound up, it could sing one of the tunes that the real nightingale sang. The Emperor's daughter was delighted.

3.1 Instrumental sophistication

Scientific research has always been a highly technical activity. Scientists have always employed the most advanced technologies available to them. The scientific revolutions of the seventeenth century were only made possible by such recent inventions as the airpump, the microscope and the telescope. It is simply not true that the best research always used to be done with ramshackle apparatus consisting of bits of glass tubing held together with sealing wax and string. A new piece of equipment often has to be improvised or specially designed for a novel experiment, but scientific instruments in general have always been at the very forefront of the state of their art. Throughout history, major scientific discoveries can be traced back to the use of new or improved technological capabilities – a more refined optical glass, more accurate machine tools, purer chemical compounds, better defined breeding stock – many of which had been developed outside the research world.

The striking feature of our own times is the extent to which scientific advances fuel general technological progress. But general technological progress, in its turn, fuels the development of more powerful techniques of research, thus providing the means for further scientific advances. This cyclic process within the science/technology complex is not merely self-sustaining: it has begun to spiral wildly outwards in scope and scale.

The most obvious evidence of this divergence is the rise and rise of 'Big Science'. In some fields of science it is now quite impossible to do serious experimental research without access to a very large, very elaborate and very expensive piece of apparatus, such as a radio telescope, a synchrotron radiation source, or a deep sea submersible. The actual research done with the aid of such a 'facility' may be carried out by many small research teams, but it is still indispensable for the further progress of the subject.

For example, the whole new discipline of planetary science depends upon the design, construction and launching of immensely complicated space probes to explore the solar system. The rapidly evolving technologies of space travel and telecommunications have not only made such research feasible: they are themselves driven to new levels by the challenges of scientific instrumentation and observation. These technical developments have made possible the remarkable scientific results obtained from unmanned missions to Venus, Mars, Saturn, Jupiter, and beyond. The success of these missions has whetted the appetites of researchers for even more elaborate, sensitive, and costly instruments. Scientific enthusiasm for the plan to put into orbit a manned space station, costing tens of billions of dollars, is only the latest outcome of this synergy between what can be conceived and what can be done.

But the development of ever more elaborate apparatus is not confined to the 'biggest' sciences. In almost every field of experimental or observational research, extremely complex instruments are now commonplace. The chemical laboratory must have its spectrophotometers, its mass spectrographs, its nuclear magnetic resonance apparatus. The cell biology laboratory must have its electron microscopes, its ultracentrifuges, its DNA sequencers. The archaeologists need access to facilities for thermoluminescence and radio-carbon dating. The sociologists must have computers to handle the data they collect – and so on.

The increasing *sophistication* of scientific instrumentation is partly due to the discovery of new ways of observing phenomena and measuring their parameters. Thus, for example, a whole range of new instruments has been developed in recent years to study the properties of solid surfaces, using soft X-rays, electron scattering and emission, ion beam scattering, topographic scanning, and many other physical properties. There is also an evolutionary factor, in that standard instrumental techniques, such as optical spectroscopy or electron microscopy, are being continually refined and developed to much higher levels of precision, resolving power, ease of manipulation, etc.

But the really revolutionary factor is the application of microelectronics to research apparatus. Scientific instruments are typical high-technology devices, in that many of their operations follow elaborate but strictly prescribed protocols. The effect to be observed often appears in a form for which peculiarly sensitive solid state detectors, such as charge-coupled devices, are now available. The signals from the detector then have to go through many stages of amplification and mathematical transformation to produce an intelligible output, in the form, perhaps, of a map or graph or table of numerical data. In other words, they are almost ideal subjects for 'silicon technology'.

There is much more to this process than installing a few microchips, digitalizing the control circuits, and computerizing the various stages of sample manipulation and data processing. Highly automated instruments are not only designed to save routine technical labour, permitting a clinical medical laboratory, for example, to attain a much higher throughput of specimens. They can also be made much more sensitive and more stable than the hand-operated instruments they replace. The work of engineering these virtues into the hardware and software may increase the price of a standard research instrument by a factor of ten, but can pay off by a factor of a hundred in both the quantity *and* the quality of the results.

It is not only that a novel technique such as electron microscopy can open up realms of observation that were previously quite inaccessible. Even a conventional technique such as optical microscopy may be transformed out of all recognition, and made immensely more powerful and sensitive, by the systematic application of advanced technological design. Quite new insights can be achieved by automating the capacity to gather and process vast quantities of data, as in epidemiology and economics. Theoretical ideas on the behaviour of a complex system, in ecology for example, can be both tested and inspired by computer simulation. In good hands, the performance and productivity of a high-technology research instrument more than justifies its sophistication and cost.

3.2 Competitive standards of instrumentation

The situation now in most fields of the natural sciences is that it is very difficult to devise a serious research project that does not require the use of one or more complex and expensive instruments. This applies even in the social sciences and humanities, where very elaborate statistical and survey 'instruments' are often required to deal with certain types of problem.

Senior scientists often bewail the increasing sophistication of the apparatus that their junior colleagues say that they need, insisting that its complexity will get in the way of their understanding of what is going on, and that it would not be needed if only they stood back and thought more deeply about how they should tackle the problem, and so on. What they are really objecting to is technological novelty for its own sake. Brute force has never been a substitute for technical brains and conceptual inspiration in the instrumentation of research.

Nevertheless, there can be no return to the situation where the individual scientist, aided by a few skilled technicians, could personally devise, and have constructed, the bulk of the apparatus that was needed for research. The skills that would now be needed are too specialized and too diverse. The virtuosos in the working of metal and glass who built the ingenious devices with which Ernest Rutherford and his colleagues explored the atomic nucleus could learn to shape new materials such as single-crystal semiconductors or brittle superconducting wires. But they could not be expected to master the whole range of micro-electronic, optical, electro-optical, electromechanical, cryogenic, vacuum, microwave (etc.) technologies that now go into the apparatus for physics research – quite apart from the software required to control this apparatus and record its output. The full range of this expertise can only be tapped by buying in many of the components of the apparatus as ready made 'black boxes', or by calling in outside specialists, as consultants or subcontractors, to deal with particular problems.

Scientific work has thus become more dependent than ever on a distinctive activity of conceiving, designing, manufacturing and marketing research apparatus. The scientific instrument industry has boomed along with the science it serves. The scientific papers in the learned journals are interleaved with glossy advertisements for all the paraphernalia of the experimental laboratory. White enamel has replaced the gleaming brass of the bacteriologist's microscope, now equipped with, say, a fully engineered mechanical system for the manipulation and micro-injection of individual cells. Digital readouts and printouts, and graph plotters, have replaced the delicately engraved vernier scales of the bench spectrograph. Rival firms extol the virtues of their off-the-peg systems for assaying, or sequencing, or synthesizing, or reproducing the amino acids in a protein or the bases in a fragment of DNA.

In the past, a research scientist embarking on a particular programme of research could collect or construct the necessary equipment in the spirit of an individual craftsman, or the proprietor of a small workshop, knowing that the quality of his or her work would mainly be limited by the skilful and imaginative use of the immediately available resources of technical labour and standard tools, augmented by the occasional purchase of a new machine to fulfil a special function.

Nowadays, planning a new research programme is much more like deciding what machinery to order, or have built, when setting up a factory to manufacture a new product line in a highly innovative industry. Should one go for the optimum performance of a specialized function, or for a less sensitive instrument with a wider range of applications? Are all the accessories offered by the manufacturer really going to be useful? Have we given enough attention to maintenance and servicing? Does this firm have a reputation for reliable products? Will we be able to interface the output with the data-processing package on our computer? Shouldn't we wait until we can get that fantastic new device that XYZ Inc. say they are just about to put on the market?

The procurement of research apparatus is thus subject to all the difficulties and uncertainties of capital investment in an advanced industry undergoing rapid technological change. These difficulties are compounded by the lack of any realistic financial criteria by which to evaluate the prospective output. The language of accountancy cannot cope with the situation. A new instrument may look immensely productive in scientific terms, but it is almost impossible to put a cash value on the research results that it might help to produce: the return on the investment, in economic terms, can only be conjectured.

Nor does the nature of the research problem necessarily determine the level of instrumentation required to tackle it. As anyone buying a new camera, or hi-fi system, or word processor quickly discovers, there are many different products, at somewhat different prices, that will do most of the things one wants in slightly different ways. In these circumstances, it is natural for the scientists themselves to argue that only the best is good enough. They know very well that they could continue to formulate and solve interesting and useful research questions with their old equipment or relatively cheap replacements. Their fear is that this work would ultimately prove fruitless, simply because their results would already have been obtained elsewhere, using more modern and expensive instruments of much greater power.

The large and aggressive commercial market in scientific instruments generates considerable pressure for ever-increasing instrumental sophistication. But as we shall see in chapter 8, the real driving force is the fiercely competitive international market in scientific discovery, linked to the even more competitive international markets in advanced technological innovation. Although these markets are very susceptible to fads and fashions, they set the standards for research instrumentation.

In principle, research results ought to be judged by the importance of the questions they tackle and the persuasiveness of the answers they propose. In practice, this judgement is strongly influenced by conventional views on the quality of the methods by which those results were obtained. Generally speaking, little attention is now given to work that is not thought to satisfy current international standards of research technology. These standards are set by the most affluent countries or corporations, able to provide their researchers with the most advanced apparatus and facilities. Technical and intellectual virtuosity may be able to compensate to some extent for limited instrumental resources, but only within a narrow range behind the rapidly advancing front line of 'state-of-the-art' technology.

3.3 Instrumental obsolescence

Research apparatus has not only become very elaborate and expensive: it also goes out of date very quickly. The scientific instrument industry matches the computer industry in the rate of obsolescence of its products. A system that was at the forefront of the art when it was purchased is superseded within two or three years by new designs. This is a not just a matter of superficial stylistic changes or minor improvements in performance, as in the automobile industry. A factor of ten improvement in accuracy, sensitivity or productivity might not be unusual.

Scientific instruments are under intense innovative pressure, simply because they are required to work under close technical scrutiny in experimental situations that have never previously been explored. Thus, for example, an infra-red detector originally designed for military use is put into an astronomical telescope. Is it sufficiently sensitive to pick up the distant galaxies they want to observe? Why is the output unstable? What factors are limiting its resolving power? Could the electronic circuitry be improved? Before long, the technological development of the instrument becomes part of the research process, and proceeds rapidly in response to the technical expertise that is lavished on it.

In other words, the research technology of any field of science is closely linked with its cognitive development, and evolves in parallel with it. Novel instruments not only make new discoveries possible: every scientific discovery also has a potential for exploitation as an instrument for further advance. Scientific instruments become obsolete just as rapidly as the scientific data they were designed to measure and the scientific understanding that those designs embodied.

Scientific instruments are a bit like high-technology weapons, in that they compete with one another on absolute performance rather than on prime cost or ease of use. In times of comparative peace – that is, in the intervals between scientific revolutions – standard equipment can be kept in use, or refitted to meet changing needs. But when the action hots up, even specially designed experimental apparatus may not survive long under the stress of serious intellectual and technical combat. And, unlike obsolete weapons, obsolete scientific instruments cannot be flogged off to guerilla scientists in Third World countries. The global scientific marketplace puts as little value on second-hand apparatus as it does on second-hand ideas.

Rapid technological obsolescence is a particularly serious problem in 'Big Science'. A large and expensive research facility, such as a nuclear reactor or fusion device, may be completed and officially opened for use after many years of construction. Yet it may soon be run down to a low level of operation, or completely abandoned, because the experiments for which it was designed are no longer thought worth doing, or can be done much more effectively in some other way.

This problem affects other very big and indivisible projects, such as nuclear power plants, weapons systems, aircraft, etc. The lead time for development, design and construction has grown so long that these projects are always technically obsolete before they go into operation. The peculiar difficulty with major research facilities is that the purposes for which they were originally required may have been left far behind by scientific and technological progress. It takes great courage to scrap a scientific

White Elephant. Sometimes it is worth spending considerable sums to adapt or retrofit it for a new purpose – for example, the Daresbury electron accelerator, which was intended for work on nuclear physics, was rebuilt as a source of X-rays for research on materials. But there is always the temptation to yield to instrumental inertia and to carry on with the original research programme far beyond the bounds of real scientific interest.

3.4 The infrastructure of research

The laboratory as a special workplace for science goes back to the seventeenth century. Many of the world's most famous research organizations – the Royal Greenwich Observatory, the Cavendish Laboratory, the Institut Pasteur, the Bell Telephone Laboratories, the Tata Institute for Fundamental Research – are still strongly identified with the special buildings originally provided to house them. Structures such as these – and innumerable others, crammed together now on overcrowded campuses – speak not only of science as a fitting subject for ostentatious patronage. They are also evidence that research has always required more than well-motivated researchers and sensitive instruments.

The modern research laboratory has to match the sophistication of the apparatus that it contains. It may have to be specially designed and constructed against every conceivable risk from highly toxic chemicals or dangerous micro-organisms. It may have to maintain an internal environment free of dust or mechanical vibration. In any case, it will have to provide a variety of services to each research bench: electrical power at various voltages and frequencies, microwaves, cold water, hot water, steam, oxygen, hydrogen, liquid helium, or whatever else is likely to be needed in regular supply. The floors must be strong enough to carry magnets weighing many tonnes, and there must be unloading bays, storage space and powerful elevators for other heavy equipment.

Instruments will have to be serviced. Mechanical and electronic workshops will be required, where special apparatus can be designed and constructed. The researchers will need a library where they can quickly consult the latest books and journals on their subject. The building will be dotted with computer terminals, linked to databases and data-processing systems within the laboratory, or elsewhere on the campus, or somewhere on an international network. Ordinary office equipment, such as telephones, photocopiers, word processors, laser printers, fax terminals, etc. will also be required so that researchers are able to communicate, formally and informally, with their colleagues and rivals throughout the world.

What we are now seeing, in effect, is a growing elaboration of the *infrastructure* of research. To some extent, this is capital intensive, as proportionately larger sums are spent on specialized buildings and on communication networks. But laboratory science now depends on a diversity of specialized support staff, comprising highly skilled engineers and instrument makers, computer programmers, information specialists, secretaries, accountants, and other office workers, not to mention the administrative staffs required to deal with the paper work engendered by this complex of activities. Many functions that used to be performed by the scientists themselves – especially when they were lowly graduate students or junior research assistants – have become so specialized and time-consuming that it is much more efficient to employ professional staffs to do them.

Once again, there is no going back to the more primitive arrangements of 30 or 40 years ago. Only a scientist who is already of Nobel Prize class can now get away with blowing his own glassware, preparing his own standard reagents, answering his letters (belatedly!) in longhand and composing his papers on a portable typewriter at a tiny desk in an odd corner of a crowded laboratory. Some of the organizational sophistication of the modern research laboratory is superficial gloss, but any scientist who has had to work for a while in a developing country can testify to the value of skilled technicians, instant telephonic communications, access to the latest publications, and all the other facilities that are now taken for granted in advanced research institutions. As the desperate state of science in post-socialist countries vividly demonstrates, these facilities are not luxuries:

they are absolutely essential for competitive research performance in the modern world.

3.5 Increasing resources required for competitive research

Scientific and technological progress is not merely the outcome of past research: it continually raises the level of resources required for further research. Everyday research equipment and supplies, specially designed experimental apparatus, large instrumental facilities, laboratory buildings, technical services and other infrastructural items, all get more elaborate and more expensive. In spite of all our time-saving techniques and labour-saving devices, the sheer cost of producing a recognizable scientific discovery or technological invention steadily increases.

This is not a new phenomenon. Over the centuries, scientists have often been hard put to it to raise the funds they needed for their research. Scientific biographies of the seventeenth and eighteenth century record heroic instances of scraping and saving, of the solicitation of patronage or of inspired improvisation, or they note the inestimable advantage of inheriting the personal wealth required to equip and run a laboratory or observatory. Many nineteenth-century institutions owe their continued existence to handsome endowments that cover their equipment and running costs as well as providing buildings and stipends. The politics of a nineteenth-century scientific community often revolves around lobbying the government for a grant of money for a research expedition, or for a large instrument, or even for the publication of the scientific record of a previous expedition!

Nevertheless, until the present century, the main resource requirement for research or invention was the personal time of the researcher or inventor. A well-to-do amateur with an independent income, such as Henry Cavendish or Charles Darwin, could not only provide himself with a workplace, apparatus and technical services out of his own pocket: he could also devote his days completely to research. When it became customary for all university teachers to be active in research, they were not heavily loaded with 'student contact hours' or administrative responsibilities, and usually had access through their institutions to laboratory space, and modest funds for equipment and assistants. The salaries of professional scientists employed on a full-time basis dominated the annual budgets of industrial and governmental research establishments, such as Thomas Edison's laboratory at Menlo Park, or the National Physical Laboratory at Teddington.

What we find nowadays is that the total cost of doing research usually amounts to several times the sum of the basic stipends of the researchers. The actual multiplier varies, of course, from discipline to discipline, and from field to field. Fifteen years ago, industrial managers were already putting it at three or more. Industrial R&D is usually much more expensive, and more lavishly equipped, than academic science, but this multiplier may well have doubled since this particular estimate was made.

In the university sector, it is difficult to disentangle research uses from teaching uses, for buildings, libraries, workshops and other infrastructural items. Only notional figures can be suggested for the way that academics divide their time between these two activities. But to give an idea of how such a calculation might come out, university finance officers now estimate that they ought to be charging something like 100 per cent extra to cover the institutional overheads incurred when they accept a research grant or contract. In other words, the indirect cost of carrying out a research project already exceeds the sum of the salaries of the research scientists, technicians and secretaries directly employed in doing the work, in addition to whatever may be needed in the way of special equipment and services.

For all the reasons outlined earlier [§3.2], the unit costs of doing research have risen, and will go on rising, inexorably. One sure consequence of scientific and technological progress is to raise the financial threshold for further progress. When budgets are limited, the question of how to maintain a competitive level of performance in a number of fields becomes the fundamental issue in a research organization.

3.6 Spreading the load

The surging costs of all the items of apparatus that a scientist now requires for even the simplest piece of research have not yet priced science out of existence. This would certainly have happened many years ago if every researcher had continued to work as an individual, with sole use of all the equipment – the microscope and microtome, the test-tubes and petri dishes, the rack of chemical reagents, the incubator, the slide rule, etc. – on his or her stretch of laboratory bench. The modern equivalents of these instruments – a scanning transmission electron microscope, an ultracentrifuge, an automatic gas chromatograph, a minicomputer, and so on – now cost hundreds of thousand of pounds apiece. The total value of the apparatus used in producing the results reported in a typical scientific paper, even in a 'little science' field such as molecular biology, could easily be several million pounds.

Once these costs became significant, they could only be contained by spreading the use of sophisticated scientific instruments over many researchers and research projects [§8.4]. The ancient tradition of providing a group of scholars with an institutional library has had to be extended far beyond the provision of general infrastructural facilities such as darkrooms, animal houses, and workshops. Many of the actual instruments of experiment and observation are only available now as communal resources shared by many researchers on a day-to-day basis.

The salient example of this development is in high-energy particle physics, where a new particle accelerator now costs hundreds of millions of pounds to construct, and similar sums to keep running. It is instructive to look back to the 1950s, when a number of universities in the UK, the USA, Western Europe and the Soviet Union were each trying to build a cyclotron, or synchrotron or linear accelerator for their own nuclear physicists. It soon became obvious that it was not possible to justify the financial, scientific and administrative capital that would be tied up in any one of these 'Big Machines' unless it was also used on a regular basis by researchers from many other institutions. The government agencies that were providing the funds insisted that they should be funnelled into a few national or regional laboratories specially set up to provide this 'facility' for a whole group of universities. Thus, in the early 1960s, almost all the UK national effort in this field was concentrated into building a proton synchrotron at the Rutherford Laboratory (near Oxford) and an electron synchrotron at Daresbury (between Manchester and Liverpool).

Nevertheless, in spite of taking a significant fraction of the national budget for basic science, these new British instruments were never fully competitive, scientifically, with what was by then available to American physicists. By the early 1970s the whole British effort in high-energy particle physics had been switched to the support of research at CERN (the European Council for Nuclear Research, whose French acronym actually means 'nucleus'), which had been established in 1952 [§8.6] and had by then built one of the most powerful accelerators in the world. Since the UK was already a major partner in this highly successful intergovernmental organization, this was a natural decision from a financial point of view. Nevertheless, it caused considerable perturbation among those British physicists who had not yet become accustomed to travelling to Geneva and working there for months at a time in close collaboration with colleagues from all over Continental Europe.

For some 20 years this arrangement has worked very well. With all the resources of European science behind it, CERN has gone on to build yet another, even more powerful accelerator, which has kept it at the forefront of research in this field. The only other comparable instrument is at Fermilab, near Chicago, where the resources of American science are similarly concentrated and shared by large numbers of physicists. But even this duopoly is now under strain. European governments, led by the UK, are questioning the cost of going on to a third-generation machine at CERN, while there is no certainty that the US Congress will continue to provide the billions of dollars that are still required to complete the rival American machine: the 'Superconducting SuperCollider'. The next step can only be the formation of an intercontinental consortium to build a 'Very Big Accelerator', to be shared by all the high-energy particle physicists in the world.

High-energy particle physics is the outstanding example of the capital intensification of research. It not only necessitates an indivisible investment of hundreds of millions of dollars or Swiss francs just to generate the primary beam of elementary particles at the requisite energy. Each separate experiment also costs millions for extra instruments, specially designed detectors, civil engineering work, electronic circuitry, computer software, etc., not to mention a year or so of the time of hundreds of qualified scientists, engineers and technicians. The only way of bringing together all the additional resources required to produce just one new scientific paper is to pool the efforts and research funds of all the experimental particle physicists in a dozen or so universities – often, as at CERN, from several different countries.

A Big Machine may also be needed just to do a Small Experiment. Much of our current understanding of the properties of crystalline and glassy materials, of liquids, of complex chemical reactions, of minerals, of biological structures, and of many other systems, now comes from observations of their effects on very intense beams of X-rays or neutrons, which can only be obtained from specially designed 'sources', each costing tens or hundreds of millions of pounds. Such a source can, however, be tapped for a large number of independent experiments running simultaneously. Thus, the capital and running costs of an instrument such as the synchrotron radiation source at Daresbury, or the high-flux neutron source at Grenoble, are spread over the hundreds of small research groups, from many different establishments and many different branches of science, who use it each year.

In astrophysics, the evolution of communal instruments has gone one step further. Large, elaborate, expensive telescopes, run by international consortia, are sited on high mountains in isolated corners of the globe. But the research scientists in the numerous university groups that study the heavens through these instruments do not have to travel half way round the earth to make their observations. Advanced communication links enable them to do their research by remote control. Sitting at their computer consoles in their home laboratories, they point the telescope at a particular star or galaxy and instruct it to transmit back to them the electronic image or other data they are seeking. A substantial investment in a communications infrastructure thus pays off both by reducing many other research costs and by permitting the use of the instrument to be shared more widely. This applies also to the great communal instruments of the social sciences and humanities, such as archives of social data, which can now be accessed, on-line, by individual researchers studying a wide variety of quite separate problems.

In the 1960s, when a university decided to provide itself with an electronic computer for research, this had to be located in a central building of the campus where users from a number of departments could come personally to take their turn at running their programs. Now, of course, immensely powerful number-crunching and database facilities can be shared by thousands of 'users', without regard to their geographical location. As a consequence, researchers are rapidly learning the art of 'networking', whereby scientists at several distant sites are able to collaborate actively on a single research project on a day-to-day basis.

The systematic use of fax, bulletin boards, electronic mail, computer networks, centralized data archives, etc. to facilitate collaborative research is now spreading to all fields of science, including the social sciences and the humanities. Direct expenditure on terminals and line rentals is more than offset by the added value of the information, expertise and instrumental resources that can thus be shared. On the other hand, the use of these facilities may be inhibited by sharpened competition between research groups in different institutions, which makes people reluctant to share their intellectual capital and technical resources.

Networking will undoubtedly become more and more customary in all fields of science; but it cannot solve the fundamental problem of providing immediate access to the diverse instruments that may be needed locally for a single piece of research. A research establishment or university department is not viable unless it can satisfy this condition. Even where a very large and indivisible 'Big Machine' is not required, the collection of instruments needed for a 'well-found laboratory' in most fields of experimental science represents a substantial investment, which can only be justified if it can be spread over a large number of research projects, being undertaken by a large number of people.

It is this condition, more than any other, that determines the 'critical mass' of concentrated effort required for effective research in any particular field. This has become a very influential concept in science policy [§6.8, §9.5], although it needs to be treated with some caution. For example, it obviously depends enormously on the actual field of study. On the one hand, in some branches of the physical sciences and engineering, a university department with a hundred academic staff may not really be large enough to justify the instrumentation required to compete directly with government or industrial laboratories in the same field. On the other hand, a pure mathematician, an anthropologist, or a social historian may only need access to a good library, a computer network and/or personal travel funds to do splendid research.

It also depends on the intellectual style that is to be adopted. This comes out strongly at the lower end of the scale, where the vague notion of a threshold of viability for a university department or self-contained research institute has traditionally been based on much less tangible considerations, such as the need to cover the various sub-disciplines of an undergraduate or graduate curriculum, or to maintain continuity of teaching activities, or to ensure a critical diversity of views and expertise, or simply to provide the mutual comforts of professional scholarly collegiality. More and more, these considerations are being overtaken by the overall trend towards multidisciplinary team research, to which we now turn.

3.7 Team research

Scientific and technological progress engenders teamwork in research. Highly orchestrated and closely managed teams of scientists and engineers are, of course, quite normal in the development, design and demonstration phases of industrial 'RDD& D'. Some fields of Big Science – notably high-energy particle physics – are also notorious for the way that hundreds of research scientists have to work together for years in order to carry out a single experiment. Many hands, eyes and brains are needed to make light of an enormous task.

But the growth of team research in almost every field of science and technology is not due entirely to the increasing scale of research projects. The advance of knowledge has come to depend on the active collaboration of scientists with specialized skills drawn from a number of distinct research areas or traditions [§5.10]. This sort of collaboration has always been necessary in the systematic attack on urgent practical problems, especially in industry. Contemporary technological and social issues seldom define themselves neatly within the traditional academic boundaries. If you want to understand the environmental effects of acid rain, you need to know a great deal about particular parts of chemistry, meteorology, geology, botany and zoology – not to mention mechanical and civil engineering, agriculture, forestry. economics, and the diverse sub-disciplines housed together as geography. The research that needs to be done cannot be left to the efforts of individual experts in each of these fields, working independently of one another.

The economic and political forces that pull basic science towards greater involvement in practical problems inevitably shape the organization of research into multidisciplinary teams – often grouped into special 'centres' or 'institutes' devoted to particular problem areas, such as 'materials' or 'social policy'. But the same multidisciplinary approach is now being used to tackle more fundamental scientific problems, whether or not they are of 'strategic' interest. Research on the basic mechanisms of memory, for example, may require neurophysiologists, molecular biologists, cognitive psychologists and computer scientists to work together as a single team. Research in archaeology demands close collaboration between experts on the physical dating of artefacts and experts on their cultural significance. Even the study of science as a human phenomenon calls for active teamwork between philosophers, sociologists and historians.

Multidisciplinary team research is thus a manifestation of the increasing 'connectedness' of the whole corpus of scientific and technological knowledge, to which we have already referred in chapter 2. It is becoming more and more difficult to formulate a question and think of a way of solving it within an isolated research tradition. It is no longer fruitful to frame each question very closely by deliberately ignoring the linkages between nominally distinct problem areas, distinct experimental techniques, distinct data sets, distinct conceptual schemes, or distinct aspects of the object of investigation. The most natural way of exploiting these linkages is to put the research into the hands of a closely interacting group of people, each of whom can look at it from a different point of view and contribute his or her particular expertise to the common pool of effort.

3.8 Interdisciplinarity

It is easy to argue, as above, for a multidisciplinary approach to every scientific and technological problem: it is not so easy to orchestrate the contributions of research specialists from very different scientific backgrounds. Effective team research involves day-to-day intellectual and technical collaboration between people brought up in very different research traditions. It cuts right across the well established frontiers between recognized academic disciplines [§7.13].

Scholars are very conscious and respectful of the features that distinguish their own discipline or sub-discipline from all others. Yet these distinctions often appear mysteriously arbitrary and inconsistent. What, pray, differentiates physical chemistry from chemical physics? How is it that research on the mechanism of vision is done in the Department of Zoology at Cambridge, in the Department of Anatomy at Bristol, and in the Department of Physiology at Oxford? At what point would one be crossing over from the history of science to its philosophy, its sociology, its politics, or its economics?

To the outsider, these boundaries between neighbouring disciplines often seem no more than conventional lines of demarcation between historically distinct scholarly traditions, rationalized by reference to somewhat nebulous intellectual criteria but designed primarily for the internal organization of academic institutions. Nevertheless, they perform an important function, and are not just withering away.

The concept of a discipline links the cognitive structure of science directly with its social structure. It defines a more or less coherent area on the map of knowledge where a group of scientists can feel at home, and identify themselves in relation to their fellow scientists. It provides a sheltered environment for researchers to learn their trade and practise their skills. Science is strengthened by the existence of social groups pledged to the advancement of particular bodies of knowledge, involving well-tried techniques, well-formed intellectual standards and characteristic problems for research.

Unfortunately, it can also happen that a scientific discipline or sub-discipline is taken over by a particular scientific 'paradigm', which then becomes an obstacle to further scientific or technological change. The rejection of the idea of 'Continental Drift' for nearly half a century by the geological Establishment shows how formidable such a social phenomenon can be. The increasing pace and mutual dependence of progress in all fields of science requires that all such barriers be kept much lower than they used to be. A botanist, for example, should not feel that she has somehow betrayed her discipline by going into toxicology, nor should a geographer be inhibited from doing research in, say, social psychology, if that is the way that his studies lead him.

A multidisciplinary team working on a well-defined research project often turns out to be an effective means for breaking down such barriers. It provides an organizational matrix, with a specific objective, for the active collaboration of people who would otherwise feel that they had few interests in common. As science advances, however, certain multidisciplinary groupings may become established as the nuclei of new disciplines in their own right. Molecular biology, for example, is now regarded as a distinct discipline, combining specific elements of biochemistry, physiology, microbiology, crystallography and physics.

The emergence of new disciplines is the central theme of academic history. The institutional maps have to be continually redrawn as knowledge progresses. Biochemistry itself only became a recognized discipline in the early years of the twentieth century, when it differentiated itself from chemistry, physiology and pharmacology. Not so long ago (admittedly, at one of the most ancient English universities!), sociology was still considered an 'interdisciplinary research area', to be run by a committee representative of the Departments of History, Economics and Anthropology. We have recently seen the creation of Departments of Cognitive Science, signalling the appearance of another new constellation of research skills and traditions.

Such developments are characteristic of contemporary science, and contribute markedly to its vitality. The idea of fostering them through 'Interdisciplinary Research Centres' goes with the grain of scientific and technological progress. Intellectual evolution can be forced along by deliberately breeding for fruitful genetic recombinations. But the equally natural evolutionary tendency towards organizational speciation and segregation still continues, and should not be ignored. Interdisciplinary groupings can easily degenerate into closed specialties with just those cognitive and institutional rigidities that multidisciplinary research is designed to overcome.

Multidisciplinary team research thus creates both opportunities and difficulties. By transcending the cognitive boundaries of existing disciplines, and looking at a scientific or technological problem from a number of different points of view, it can bring unprecedented intellectual and technical power to its solution. But it is not a simple matter to arrange, and generates significant changes in the way that scientists are trained and employed, in their professional commitments, and in the practical organization of the work of research.

3.9 Specialization and collectivization

One of the main effects of scientific and technological progress is thus to hasten an overall trend from individual to collective modes of work. This trend goes back a long way. By the middle of the last century, several German universities had already set up what we would now call 'graduate schools': groups of young scientists being trained as researchers by collaborating in the research projects of established professors.

This relatively informal mode of collective effort has spread throughout the academic world. It is quite obvious that the 'school' of a famous scientist such as Niels Bohr, Peter Medawar or John Maynard Keynes contributes much more to knowledge than the sum of the labours of its individual members. Inspired and guided by such a leader, a whole group of graduate students, post-doctoral fellows and junior faculty members may pursue a single line of investigation together, and collaborate much more closely than would be apparent from their scientific publications.

Until recently, each member of such a group would normally expect to take personal responsibility for the outcome of his or her particular research projects, even when local customs and courtesies prescribed that their papers should also bear the name of 'the Prof' as a co-author. In the early 1960s, however, Derek de Solla Price drew attention to the increasing proportion of scientific papers with more than two authors, indicating that they had been working together as a team. In many fields of science it has now become unusual, even for experienced researchers, to work on their own, following their own bent in the choice and performance of their research projects.

The strongly individualistic tradition in research is rapidly being superseded. This form of 'collectivization' may even be accelerating. A few years ago, in a typical field of 'little science' such as molecular biology, a group of three or four researchers with a common disciplinary background could collaborate scientifically and make a serious contribution to knowledge with quite modest resources. Now it seems that the only way to achieve a major advance is to concentrate and combine the efforts of a tightly organized team of a dozen or so specialists, with carefully chosen but diverse skills and a battery of instrumental facilities at their command.

The trend is also towards *specialization* in the problem areas chosen for research or development. Faced with so many more scientific and technological opportunities than resources to explore or exploit them, research units, academic institutions – even whole countries – find that they must concentrate these resources quite narrowly in order to make progress against their competitors in the world markets of knowledge and industrial products. The cost of research has not only to be shared by the provision of collective facilities: it has to be kept down by a formal or informal division of the labour of research, nationally and internationally, into highly specific areas of knowledge, technique, or commercial application.

This pattern of specialization, once it is established, becomes more firmly entrenched through competition. A large research group working in a narrow problem area is in a position to pick up bright ideas from other groups, and exploit their breakthroughs before they can do so themselves. Its international visibility may dazzle non-specialists, and its organizational weight may give it a larger proportion of the available national resources than it deserves. What Robert Merton dubbed the Matthew Effect – 'To him that hath shall be given' – applies to research groups as well as to individual scientists.

A collectivist ethos has thus evolved, favouring large and highly differentiated research units. The critical mass of people and instruments thought to be necessary for effective research in any particular field may not be determined solely by such objective factors as instrumental sophistication or multidisciplinarity. Even in fields where all that individual researchers really need is access to a library or a computer terminal, which could easily be provided by a communications network, advantages are now seen in bringing them together into specialized groups.

The intellectual environment in such a 'centre of excellence' can certainly be very stimulating, both for mature scientists and for the graduate students they are training in research skills. But the disbenefits of specialized research collectives should not be ignored. In spite of aspirations towards a multidisciplinary approach, the effect may actually be to reduce the diversity of expertise around the common room coffee table, where unplanned, informal contacts between specialists in widely different subjects are often so fruitful. The scientific leadership of such a group is also exceedingly demanding, and can be disastrous if performed less than excellently. Unless the scientific performance of the group is periodically evaluated, it can soon degenerate into another intellectually closed, rigid and mediocre social unit.

The process of concentration to produce such collectives also works against broad, multidisciplinary educational developments. In practice, it cuts across the traditional links between undergraduate teaching and research, which many scientists regard as essential for scientific progress. Even for graduate students, the complex division of labour in a large team working on a long-term research project may produce an over-specialized training experience.

4
Transition to a new regime

Jack saw that the beans had all come up and had grown most wonderfully.

4.1 A history of rapid, unimpeded growth

Ever since modern science 'took off' in the seventeeth century, it has been a growth industry. Knowledge and technical capabilities have not only accumulated steadily: the rate of accumulation has also accelerated over time. The scale of all scientific and technological activities has continually expanded. Every measure of these activities – numbers of people engaged, resources employed, output of published papers and patents, commercial and industrial impact, etc. – seems to have been increasing exponentially for the best part of three centuries.

Of course, that same period has seen immense growth in many other parameters of national life: population, industrial production, 'gross national product', education, and so on. Nevertheless, the overall growth rate of science throughout this whole period has been exceptional. In the 1680s, for example, England had a total population of about 5 million: nowadays, the United Kingdom has a population of about 50 million – ten times greater. Even if we multiply this by another factor of ten, to allow for the fact that people are now much richer individually than they were before the Industrial Revolution, we still fail by a whole order of magnitude to match the thousandfold growth of scientific and technological activity in Britain since that time.

Much the same elementary historical calculation could be made for other European countries such as France and Germany. In more recent years this process has extended to almost every other industrialized country in the world. We tend to be dazzled by the spectacular case of Japan, where almost all expansion has taken place in less than a century, but many other countries, including such great nations as India and Brazil, are already some way along the same path.

This general picture of modern science as an activity with a phenomenally expansive and successful history scarcely requires quantitative support. Indeed, it is not so easy to give a precise figure for the long-term historical rate of growth of any of the various indicators that are now used to compare the scientific and technological activities of different countries.

A figure that is sometimes quoted is an overall growth rate of between 5 and 7 per cent per annum. This corresponds to Derek de Solla Price's observation, in 1963, that the world total number of scientific journals had apparently doubled every 10–15 years since 1665. This is a useful figure to keep in mind, but it is subject to a number of important reservations.

In the first place, a global average is misleading, in that it conceals large variations in the rates of growth, from time to time and from country to country. Thus, for example, the scientific take-off of a large country such as the USA, the USSR or Japan may compensate for a period of scientific stagnation or decline in other countries or regions. In some respects there is now only 'one world' of science [§8.10], but national trends are still of greater social and personal significance than international averages.

The aggregated rate also conceals large variations from one field of science to another. The advance of knowledge opens up new fields which grow very fast indeed, while others mature and lose their scientific interest. The finer the scale on which scientific activity is measured, the greater the variation in the observed rate of change. Thus, for example, the total literature of a sober discipline such as geology, characteristically doubling every fifteen years, may include several exciting subfields such as marine geology which grow by a factor of 5 or 10 during the same period. The rate of change of the total amount of scientific activity is thus a poor measure of the rapidly changing nature of that activity.

Again, the output of scientific papers is a rather limited measure of scientific activity. It is not easy to think of any other quantifiable feature of science that has remained so stable over such long periods. Nevertheless, quite apart from considerable uncertainties and long-term historical variations in what are thus being counted, bibliometric indicators define 'science' too narrowly. In particular, they underestimate the many forms of scientific activity that contribute directly or indirectly to practical life, through industry, medicine, agriculture, etc.

The number of people actively engaged in science at any one time would be a useful indicator of its scale. Unfortunately, this number is difficult to define precisely, or to determine historically. Before the middle of the nineteenth century, scientific research was not a distinct profession. The number of active 'scientists' in those days can be estimated by counting their publications, but even then there were many technical craftsmen, inventors, medical practitioners, clerics, engineers, and others who were also doing essentially scientific work and who contributed by a variety of other means to the total edifice of knowledge.

In more recent years, technological research and development has expanded so rapidly that it now employs the majority of the people who might be counted as professionally engaged in 'science'. Science thus shades off into other practical activities such as engineering and medicine, not to mention the teaching of science in universities, polytechnics and secondary schools. We may be sure that the proportion of the population thus employed is very much larger than it used to be, but quantitative historical trends or international comparisons of aggregate statistics covering such heterogeneous and ill-defined groups are not very meaningful.

Scientific activity is often measured nowadays in terms of the amount of money spent on it. Most governments indicate their expenditures on science in their national budgets, although with differing conventions from country to country [§5.7]. Thus, for example, something like half the funds supposedly devoted to academic research in the UK is represented in these figures by a purely notional proportion of the total government grant to the universities. In France, by contrast, these funds flow through separate channels and can be accounted for in detail. International comparisons are subject to all the uncertainties of exchange rates, while estimates of historical trends over long periods are confounded by the difficulty of making allowances for inflation.

In any case, the large-scale funding of research by national governments is a relatively new phenomenon. Before, say, 1918, only a very small proportion of scientific activity was actually supported from the public purse. The other sources of support – academic, commercial, charitable, or simply personal – were so fragmented and diverse that it is very difficult to account for them all and to aggregate them in financial terms. Even now, governments do not directly fund all scientific activity. In a developed country such as Britain, something like half the total national expenditure on scientific and technological research is reckoned to take place in the private sector. But many companies (at least in the UK) do not publish the amounts they spend on R&D, and standard accountancy procedures tend to lump together expenditures on basic research, applied science, technological development, product design and pre-market testing.

Any assessment of the total national expenditure on science thus has to include some very uncertain figures, such as a notional proportion of the salaries of university teachers and arbitrarily defined estimates for the R&D expenditures of private firms. The standard public data for the major industrial nations since 1945 can be used (very cautiously!) for contemporary international comparisons, but they cannot be extrapolated back over the centuries to give meaningful measures of long-term trends in total scientific activity.

Having said all this, we are still left with overwhelming evidence that Western science has a history of continuous expansion in all its significant parameters. From time to time – for example, in early Victorian England – there have been complaints of 'national decline', but these usually referred to its supposed quality or economic impact rather than to any numerical measure of its scale. Indeed, if the bibliometric data are to be believed, whenever scientific activity was curtailed or diverted during a major war, the lost output was made up again shortly afterwards, as if the curve of exponential growth were simply jumping over a small ditch in its upward path. This immemorial tradition of going on from strength to strength is a key factor in our present feeling of crisis.

4.2 From occasional patronage to commercial contract

Continuous expansion in every parameter has been one of the unchanging characteristics of science and technology. The philosophical principles that underpin and motivate the scientific endeavour have also lasted remarkably well since they were instituted in the seventeenth century. But many other features of science as a human enterprise have changed out of all recognition. The vocation of a loosely connected community of a few hundred 'savants' in Western Europe has grown into a major social activity involving millions of people, highly organized nationally and often operating on a world-wide basis. Indeed, no social institution could possibly have accommodated the enormous effects of three centuries of exponential growth, at an annual rate exceeding 5 per cent, without radical qualitative changes in its nature, function and structure.

Some of the changes necessitated by the advance of knowledge and of technique have already been discussed in the previous chapters. Scientific disciplines and practical technologies have evolved, split apart, proliferated, and recombined to produce a multiply connected map of knowledge whose complexity passes ordinary human understanding [§2.2, §3.8]. The techniques of research itself have become so refined and elaborated that they require the coordinated efforts of many diversely specialized hands and brains [§3.7]. These developments have necessitated the invention of many novel social practices – doctoral training, peer review, instrument design, project management, to name but a few – and the establishment of many novel social institutions – university departments, graduate schools, abstract journals, interdisciplinary research teams, Big Science facilities, and so on. Even within the short span of an individual scientific career, there have been bewildering changes in the intellectual structures, the material conditions and the social relations of laboratory life [§1.2].

The dramatic changes that have occurred inside science have been accompanied by equally dramatic changes in the relationships between science, technology and society at large. The mere fact that science has become a much larger and much more complex activity was bound to have this effect. In particular, the accelerating cost of apparatus and other facilities [§3.5] could only be met by a disproportionate flow of material and human resources into research, with a corresponding elaboration of the machinery of accountancy and management. The affairs of a relatively self-contained community have thus been opened up to a quite unprecedented degree of external scrutiny and control.

The other significant development has been the erosion of the traditional boundaries between 'pure' science, on the one hand, and 'applied' science on the other [§2.3]. It has become more and more difficult to draw a sharp line between curiosity-oriented 'research', and use-directed 'invention'. These two modes of scientific activity often co-exist at the same laboratory bench and merge imperceptibly into one another as a cognitive novelty evolves into a technological innovation. Indeed, for many purposes, it is no longer meaningful to separate 'science' from the variety of other highly technical activities – performance specification, design engineering, prototype testing, clinical trials etc. – that contribute to the overall process of 'Research and Development'.

In effect – to extend a stock metaphor – the tang of potential utility can be detected further and further upstream towards the headwaters of basic scientific discovery [§2.5]. Those who hope to gain from exploiting that promise are taking an increasing interest in stimulating basic research and steering it towards their own ends. The fact that research can apparently be targeted more and more effectively towards improvements in various material aspects of life – industry, health, national security, and so on – inevitably generates a demand that these targets should be set by the corresponding organs of society, rather than by the researchers themselves. In other words, the whole scientific enterprise is being more and more directly sponsored and programmed to meet explicit societal needs.

Historically speaking, this has been a long process, which can be traced back to the foundation of such institutions as the Royal Greenwich Observatory and the French Academy of the Sciences at the end of the seventeenth century. But the transformation of science into a more central but more dependent organ of society has accelerated since the Second World War. This is particularly apparent in the way that 'academic' research is funded. University research work in the humanities, the social sciences and medicine is still largely supported by a mixture of academic, philanthropic and state patronage, along more or less traditional lines. But in the natural sciences and engineering the brunt of the expense is now met out of the ample purse of the state, which has increasingly exercised its right to call the tunes it pays for.

This has been a relatively gradual development, world-wide in its generality, but following different paths in different countries. The whole story is much too complicated to tell in full, but is well exemplified by what has happened to academic science in the UK since the 1940s. Originally, this was just something that went on in universities, with occasional private or public patronage. Then an administrative apparatus of quasi-independent, quasi-academic 'research councils', was created, partly to support research in its own units and establishments and partly to distribute public funds to university staff to cover the *additional* costs of research projects – for example, for apparatus, technical services, and especially post-graduate students – that could not be provided by their institutions. These grants were awarded entirely according to the intrinsic scientific merits of the proposed research, and the meritocratic credentials of the students – never according to any criteria of potential use.

As research costs have increased, however, academic scientists have become more and more dependent on this type of direct support: in many fields, it is now scarcely possible to undertake serious research without a grant from a research council or other funding body. Governments have naturally tried to get better value for their money by increasingly elaborate procedures for selecting projects and accounting for the funds received. Almost inevitably, pressure has also come from political and administrative circles to give greater weight to the potential applicability of the results, as distinct from their purely scientific interest.

The whole funding system for academic science is still in a state of flux. As we shall see in the next chapter, the pressure is now on for the introduction of a new regime, supposedly more responsive to the needs of industry and commerce, and more in line with current beliefs in the power of market competition to ensure efficiency and excellence. The idea is that state support for all potentially exploitable academic science – in the extreme, this would include every item of 'strategic' research designed to contribute to a generic technology [§2.5] – ought to be phased out, and replaced by direct funding from the private sector of the economy. In pursuit of the same goal, a number of research council and government establishments doing quasi-academic research in agriculture, public health, environmental protection, etc. have been privatized, or forced to obtain the bulk of their funds through research contracts with industrial or governmental 'customers' [§6.1].

This sketch of the movement from sporadic patronage to customer–contractor relationships in the funding of academic science in the UK can be parallelled in many other countries. In some countries, such as Germany, many features of the traditional system are still maintained; in other countries, such as New Zealand, the trend has gone even further, with a systematic policy of putting all the arrangements between the government and the universities on a quasi-commercial basis. To some extent, this whole movement can be associated with a world-wide trend away from 'command' systems of management towards a more devolved 'market' orientation in every aspect of human affairs [§5.9]. But the change is also being driven by major factors within science, and in its relations with the society that supports it.

4.3 Changing professional roles and career paths

Major changes in the social relations of science are having profound effects on the way that research is organized, in universities, government laboratories and other institutions. Striking evidence of these structural effects shows up in changes in the professional roles and careers of scientists [§3.5, §4.1]. The very word 'scientist' was not invented until the middle of the nineteenth century. Before that, there were very few people paid directly to do scientific research. Most contributions to the scientific literature came from people with a variety of other callings, such as teaching, medicine or the enjoyment of private means, while many inventors were almost unlettered craftsmen.

More than a century ago, the natural sciences began to take firm root in universities. Following the German lead, it became customary for basic research to be undertaken on a part-time basis by university teachers. This way of organizing 'academic' scientific activity is now so deeply entrenched in British and American higher education that it is regarded as the norm, although other countries manage these matters somewhat differently. For example, much of the best academic science in France and Germany (and to a considerable extent, still, in the former socialist countries) is done by full-time research scientists and scholars who are not simultaneously employed as university teachers.

As a result of these and other developments, such as the rise and rise of industrial research, science rapidly became fully professionalized, and scientific roles and careers became established far outside academia. Scientists with much the same training and skills were employed as researchers and technical experts in a wide range of organizations, such as research council establishments, government laboratories and many large industrial firms.

But the increasing finalization and collectivization of science and technology [§2.3, §3.9] have made it more and more difficult to make a practical distinction between the work of 'scientists' and the work of other technical personnel who would normally call themselves 'engineers'. For example, a scientist with a PhD, recruited by an industrial firm to do strategic basic research, may soon move across into technological development, and after a spell in production engineering finish up in the marketing division of the company. At the same time, researchers are having to collaborate in multidisciplinary R&D teams [§3.7], and are thereby losing their academic identity cards. A team working on, say, the safety of a new type of nuclear reactor might include a systems analyst originally trained as a research astrophysicist, an academic nuclear physicist now concerned with instrumentation, and a chemical engineer who has become expert on theoretical hydrodynamics, all under the direction of a materials scientist who is beginning to show distinct managerial talents.

The industrial and academic worlds have also come much closer together, and interpenetrate at many more points than they used to [§2.3]. Scientific problems related to the needs of industry or defence are being studied in many university laboratories: conversely, many industrial and governmental research laboratories are working on quite fundamental scientific questions. In such cases, the question of who actually does the piping or pays the pipers may not be so significant as the fact that they are all playing the same repertoire of tunes.

Some of the most recent changes in the research system – for example, the transfer of some government and research council establishments to the private sector [§4.2] – have taken place so abruptly that they have quite bewildered the people directly affected by them. Even in universities and polytechnics, the tradition of part-time research by teaching staff with permanent academic tenure has given way to a situation where a considerable proportion of the fully qualified scientific workers are now employed as full-time researchers on short-term contracts. These fundamental changes in professional roles and expectations will be taken up in more detail in chapter 7.

4.4 The inevitability of structural change

It must be emphasized, however, that the apparently settled and tranquil patterns of life that were thus disturbed had not, in fact, been established since time immemorial. Science and technology are dynamic activities which continually transform themselves by their own successes. Radical changes in the objectives and techniques of research are nothing new, and have always given rise to new organizational structures where individuals have been faced with quite unexpected – not always unwelcome – career paths and prospects.

It would far exceed the scope of this book to give a detailed account of the long and complex history of such changes. Nevertheless, two characteristics stand out. Superficially, the first is a commonplace: that the relationship of science to society at large has continually grown closer, and more important to both parties. Science, through technology, has become integrated into the workings of our economy, our polity, and our culture, and has developed into one of the major forces for their structural transformation. What we must also realize is that science itself is transformed internally as it reacts to these forces.

Take the well-known historical phenomenon of the increasing influence of science on the technology of *war*. This has not only revolutionized every aspect of warmaking, from the clash of arms on the battlefield to the place of defence in national policies and budgets: it has also profoundly affected the way in which research is managed, far beyond the laboratories of defence departments and defence industries. The civilian science that was enlisted for the duration of each of the World Wars of our century was demobbed as a veteran, hardened and rather coarsened by the experience. Indeed, in the Second World War science had demonstrated its strength so convincingly that it was effectively retained with the colours.

The fact that defence now takes somewhere between one-quarter and one-third of all our national expenditure on R&D represents a fundamental change from past policies for science and past practices within science. It is not just a matter of politics when, for example, a number of knowledgeable and thoughtful researchers are prevented from telling publicly all they know about certain scientific matters, or when immense R&D programmes are launched and pursued without the benefit of really independent technical evaluation [§2.7]. It is wrong to suppose that 'science' is just a black box, or a set of separate black boxes, whose internal workings are unaffected by such powerful external forces.

The second general characteristic of all these changes is that interconnection and unification are as much part of the scientific tradition as proliferation and differentiation [§2.2]. In the late nineteenth century, for example, the German chemical industry established the first industrial research laboratories. This new type of institution was copied in other industries and in other countries, developing into such giant organizations as the Bell Telephone Laboratories in New Jersey, the Philips Laboratory at Eindhoven, the ICI Corporate Laboratory at Runcorn, and many others. By the 1930s, industrial R&D had evolved into a major new type of scientific activity, quite distinct from academic science in style and organization. New professional associations, such as the 'Institute of Physics', had even been founded to protect the interests of 'industrial and applied physicists' from domination by the 'pure physicists' in the 'Physical Society'. By the middle of the 1960s, however, these two associations had merged into what is now called the Institute of Physics, in line with the development of much closer technical, organizational and career links across the various sectors of the national research system.

It is natural enough, in elementary accounts of the 'internal' history of science and technology, to represent the germination and growth of new practices, new roles and new organizational forms in isolation from one another, as if responding to particular technical developments or societal demands. In reality, however, what we still call, very loosely, 'the scientific enterprise' is neither an arbitrary collection of disparate, disconnected institutions nor a closed, self-contained social entity. It is, as it has always been, a continually evolving system of closely linked elements – intellectual and institutional – many of which are themselves growing and changing in relation to one another.

4.5 Budgetary evidence of a transition to a new regime

Scientists in every past era have always taken the uninterrupted expansion of science and technology for granted. Even in times of economic depression or war, when research jobs have been in short supply or pressed into the service of the nation, this process of expansion has scarcely seemed to falter [§4.1]. A dynamical combination of continuous rapid growth, proliferation, differentiation and re-integration has always seemed the natural condition of the scientific enterprise.

This confidence in ceaseless expansion is much more than a general belief in scientific 'progress' – that is, in the accumulation of successively more valid scientific knowledge and more effective technological capabilities. It assumes that there will always be room in the future for good new ideas, good new people and good new enterprises, regardless of what happens to older ideas, older people and older institutions. In the celebrated phrase of Vannevar Bush, science has always been regarded as an 'endless frontier', where the opportunities for the persevering pioneer will never be closed.

As Price pointed out in 1963, this assumption is deeply ingrained in the whole culture of science. Although seldom openly voiced, it is one of the implicit conditions for the functioning of many of the traditional practices of science. For example, much of the day-to-day work of academic research is done by bevies of graduate students and post-doctoral assistants, undergoing training as independent researchers and research leaders. The number of people who go through such an apprenticeship would greatly exceed the number of posts that they might eventually hope to fill, unless the number of these posts were increasing very rapidly. Both at the mundane level of career opportunities, and at the psychic level of the relationship between scientific generations, this expectation is as fundamental to the scientific culture as the expectation of general population growth in ordinary family life. If there came a time when it was no longer valid, then the nature of scientific activity itself would be radically altered.

But this time had to come. Exponential growth could not outrun the resources that fed it. In 1963, national R&D expenditures in a number of economically advanced countries were already around 1 per cent of the GNP [§4.1]. If the expansion of scientific activity (measured in these terms) were to go on at the previous historical rate, then by the end of the century it would take up around 5 per cent of the national income. Extrapolate to the 2030s, and the figure would reach 20 per cent – and so on, *ad absurdum*. In other words, Price argued, this sort of growth could not possibly go on at the same high rate for more than a few decades, and must soon slow down.

Price's general argument for an eventual transition to a new regime of much slower growth was, and remains, incontrovertible. The only question was: when and how will it occur? Price thought in 1963 that science might then have reached the point of inflection on an 'S-curve', with still some way to go before it reached its 'limit to growth'. This prediction was largely ignored at the time, and yet it has actually proved more sanguine than the reality of the past 25 years.

The most obvious evidence of a break with the past is in the behaviour of the crude, nationally aggregated indicators of overall scientific activity. From 1970 to 1985, the UK has followed the general international trend in developed countries, which was for R&D expenditure to level off for a while at around 2 per cent of GNP and then to rise slowly to about 2.5 per cent. The number of scientists and engineers engaged in R&D as a proportion of the labour force has shown similarly modest growth in that period.

To appreciate the significance of this observation, recall that previous trends would have shown both these indicators doubling in that period. If the expectations of scientists in the 1960s had been realized, an industrially advanced country like Britain would now be devoting something like 5 per cent of its national income to scientific work. Whether or not this would have been desirable, or even feasible, it has certainly not happened (even in Japan), and does not seem likely to happen in the near future. Indicators of this kind are notoriously inexact [§4.1], but their presumed errors and uncertainties are much smaller in magnitude than this striking effect.

It should be said, however, that the budgetary data are too imprecise to provide evidence for a *sharp* transition to a quite different regime of 'non-growth'. The period since 1970 has seen a major world-wide recession in general economic activity, and would in any case be too short to demonstrate a definite break in a long-term historical trend. Indeed, *world* output of scientific papers is still growing steadily, as newly developed countries expand their scientific activity. The indicators for Japan and West Germany continued to increase at nearly the traditional rate, and even in the USA there was a significant upward turn during the 1980s.

The disaggregated data for the UK show, moreover, that the change of pace is not uniform over all forms of scientific activity. The transition to nearly level funding has occurred only in the university sector – and the data may underestimate increasing collaboration between academia and industry. There was substantial growth for some years in expenditure on defence-related R&D, both by government and industry. There are also some signs of a resumption of growth of general R&D funding in the private sector, with industrial firms doing more basic research in-house. An optimist could thus interpret the plateau in the overall budgetary indicators for UK science as no more than a pause, associated with Britain's difficult economic circumstances at a time of world-wide recession.

A reliable *quantitative* picture of what really has happened in recent years to the various components of the R&D systems in various countries would obviously be very helpful. Unfortunately, the picture that we have at present is extremely sketchy, if not positively misleading at some points. Some of the published national and international indicators are reasonably reliable, within their stated terms: others contain conjectural figures, or refer to ill-defined categories that do not merit comparison from country to country or from time to time. Some of the required information could probably be dug out of government statistics, etc., but other important facts may only be obtainable by very extensive and intensive research. We simply do not yet have sufficient quantitative data to corroborate or refute the various subsidiary hypotheses concerning the nature of the change that has obviously occurred.

Until this research has been done, it is more instructive to consider the *qualitative* evidence for a radical change in the polices and procedures of the many different bodies through which science is funded and managed. In spite of all the talk about intellectual property rights [§2.7], scientific knowledge is not just a commodity, subject to the iron laws of economics and measured in pounds, pence and personnel: it is also the product of a complex, highly institutionalized human activity. The future actions of most social institutions can be predicted rather more reliably from the attitudes and express intentions of their effective policymakers than from numerical extrapolations of immediate past trends.

4.6 Indications of a change of attitude

The standstill in the overall funding of R&D in the UK since the early 1980s points to a major change in the nature of scientific activity. International comparisons suggest that the UK is not untypical in this respect. There are grounds for believing that science in general is facing 'a version of limits to growth' (as an editorial in *Science* put it, referring to American science!), where its immense expansive potential is held in check much more closely than in the past.

There are no signs that the scientific enterprise has lost its way or is running out of steam [§2.1]. Scientists are as confident as ever that there are plenty more incredible discoveries they could make, if only they had the means. Technologists are equally confident of their capacity to create inventions that the world at large will value and desire. Science and technology remain intrinsically expansive, in that every advance still seems to spawn a dozen new ones [§2.4].

Scientific activity in the UK since the 1980s has clearly not been limited by a shortage of ideas for really good scientific projects or an absolute lack of the means to undertake them: it has been limited by societal reluctance to pay for them. Science is no longer cheap, and it does not offer an immediate return to an investor. The cost of research has to be met somehow, whether through taxes, reduced company dividends or increased consumer prices. There are sound general reasons for believing that money spent thus will be repaid many fold – in due course. But any activity requiring a total layout of several per cent of the national GNP is in serious competition with other desirable items of public or private expenditure, and has to be justified in that context.

There is no evidence that resistance to the further expansion of science stems directly from any single centre of authority. It would be wrong to suppose that the Cabinet or the Treasury lack all appreciation of the value of science as a long-term investment. Indeed, until a few years ago total UK expenditure on R&D was not an operational concept: it was merely a statistical aggregate of items spent independently by many separate institutions: universities, research councils, government departments, industrial firms, etc. [§4.3]. Only now, for example, are questions being asked, within and outside government, about the overall balance between civil and military R&D. The simultaneous levelling of expenditure on many of these items was not apparently the result of an executive decision directed at science as such.

What we are seeing, rather, is a major change in *general* economic and political attitudes towards the support of scientific activity. The new attitudes pervade the whole economy and polity, at all levels of responsibility. The actual change may have been triggered by the rapid worsening of the general economic climate in the mid 1970s, and accelerated by the stringent fiscal policies of the present British government. But it is not just a little local difficulty that will go away when times are better.

The most obvious evidence of this change is in the way that decisions are made about funding. When scientific activity was still expanding rapidly, such decisions were strongly 'science driven'. Technical criteria of merit often ranked above prime cost or economic benefit. Budgets for basic research were allowed to grow to meet the rapidly rising costs of instrumental sophistication, even up to the provision of 'Big Science' facilities on an unprecedented scale. In academia, in research council and government laboratories, and even in industry, the great majority of 'alpha-rated' (i.e. top quality) research projects could be supported just because they were 'good science', or because they offered a reasonable hope, in the long run, of making significant progress towards solving recognized practical problems.

Nowadays, such decisions occasion much more agonizing debate. Stricter budget limits mean that hard choices have to be made between projects of great promise. This scrutiny inevitably goes outside the frame of strictly technical considerations [§5.4]. The authorities providing the funds are asking questions that go beyond absolute assessments of scientific merit to relative evaluations of presumed practical outcomes. Project is weighed against project, programme against programme, and field against field, according to criteria that are heavily influenced by political or commercial priorities.

It can be argued that science is merely going through the same managerial revolution as most other social institutions. Like the Army, the City, the Civil Service, medicine, public utilities, local government, and education, it simply has to become 'leaner', more efficient, more accountable, more competitive, more concerned about wealth creation, and so on. In a word, science must be expected to keep up with the times. It can also be argued that the times themselves have changed in other ways, and that the general public has become disenchanted with 'socialism', and would prefer to spend its own money according to its own priorities, rather than being taxed heavily to provide the government with funds that are wasted on apparently useless research.

Trite and ill informed as they often are, such arguments are not without force. The changes that scientists perceive in their own domain are mostly particular forms of changes that are going on all around them, in every other walk of life. Our advanced industrial society is passing through a period of significant structural change – brought about, perhaps, by such techno-scientific developments as the computer – in which science and technology are being carried along by a swiftly flowing tide. But that is compatible with – indeed, makes more likely – a fundamental change in the perceived social role of science, and the forces that shape that role in practice.

These forces operate in a variety of spheres: political, fiscal, commercial, demographic, and cultural. They interact in a multitude of ways and have very diverse and complex effects. It is possible, however, that they all have a common factor: they are all affected by a significant shift in public opinion concerning the value and purpose of 'science', as distinct from medicine, engineering and other useful science-based technologies.

Scientific knowledge is certainly still valued as a cultural product in its own right [§2.6]. To judge by the books that get published, the products of research in the more academic and/or arcane branches of the social sciences and humanities are not being spurned by those sufficiently educated to appreciate them. What would have seemed in the past to be vast sums are still being spent on instruments whose sole purpose is to explore the depths of space, or to test theories for the unification in principle of the ultimate forces of nature. Public interest is as lively as ever in some of the traditional questions of 'pure' science: the origins of the human species, the significance of dreaming, the fate of past civilizations, the birth and death of the universe. Curiosity-driven research in animal behaviour, ecology, and other branches of natural history is also widely acclaimed, both for its connections with the conservation of the natural environment and for the intrinsic beauty of the visual images that it provides for television.

It is not clear, however, that the general public is inclined to put its money where its mouth is. Even when the total 'science vote' was growing, the proportion going to basic research was declining. Efforts by the Royal Society and other bodies to lobby for academic science and scholarship have not catalysed a wider public movement. The only arguments that now seem to carry any weight for the expansion of science are those that emphasize its promise of future wealth or other tangible benefits.

Perhaps this is nothing new. Society at large has never had much use for the philosophical. Public financial support for science has always had very worldly motives. Much of the immense cost of particle accelerators, for example [§3.6], has been borne patiently by governments and legislatures in the hope that the high-energy physicists will eventually discover another pot of gold like their nuclear precursors.

The change of mood that we have seen in the past decade or so is something more subtle than 'disenchantment' or 'disillusionment'. The most apt word might be 'impatience'. People outside science feel that it has been too slow to deliver the goods that they thought they had been promised, so they want to hurry things along. Many politicians seem to believe that if all the resources and technical virtuosity employed in basic science, apparently to satisfy the curiosity of the scientists, were more efficiently managed, and focussed more sharply on practical problems, then they would yield a greater return. The ultimate benefits of long-term research are thought to have been oversold, and results are now being demanded on a much shorter time-scale, primarily to deal with the urgent economic problems of the nation.

This applies particularly to the social sciences, whose nature and purpose are not widely understood, even though their subject matter is much more directly related to everyday experience than most of the natural sciences. Even quite well educated people dismiss research results in disciplines such as sociology as either 'obvious', or 'unconvincing', or else expect them to be capable of immediate and direct exploitation. After a period of enthusiasm, the ultimate social utility of such disciplines is now seriously distrusted. They are therefore under extreme pressure to demonstrate on the spot the utility of their findings – in contrast with physics or biology, where it is still generally accepted that it may take some years before a promising discovery can be developed to the stage of commercial application.

This change in public mood should not affect the policies of industrial firms, which have long been accustomed to investing heavily in R&D for the development of their existing products and processes or for longer-term technological innovation. They must be presumed to understand the wealth-creating potential of this activity, and to have their own criteria for assessing its profitability. They must surely know the dangers of trying to get away with less expenditure on R&D than their competitors, nationally or internationally. And yet many British firms – even some major multinational companies – have been running this risk for years. Is this a sign of declining public confidence in science, or is it just another manifestation of the general 'short-termism' induced by the financial arrangements and accountancy conventions of British industry?

Again, it is surprising that the same mood of impatience seems to be affecting research that is directly related to urgent practical social problems, such as public health or environmental pollution. One would normally interpret a restriction or reduction of expenditure on this type of R&D as a reduction in the political salience of the problem in question. Nevertheless, over a period when environmental issues were moving higher up the political and economic agenda, funding for the Natural Environment Research Council was being drastically cut. And how is it, to give an example from the social sciences, that basic research in criminology has not been given the urgency that current concerns about law and order would suggest?

This parsimonious attitude impacts heavily on government funding of 'science' in the narrower sense – that is, research motivated primarily by curiosity, without any applications in mind, or research whose possible outcomes are very difficult to evaluate in practical terms [§2.5]. And yet nothing has happened to justify this loss of confidence in scientific research as a long-term investment which must, on the average, eventually pay handsome returns. As we saw in chapter 2, there has not really been any reduction in the intrinsic creativity of science, whether for knowledge or for wealth. Indeed, the opportunities for the technological exploitation of its results are probably greater than ever before [§2.4].

In the end, society may be willing to allocate a substantially larger share of its financial and human resources than it does now to the more academic forms of 'strategic' research [§2.5]. But now that the resources required to support this 'science base' can no longer be regarded as a marginal item in the national budget, science will have to compete robustly with rival institutions to win a better share of the cake. And that means that it can no longer escape from the demand that it accounts convincingly for what it achieves with those resources.

4.7 Radical change within a steady state?

In just a few years, general economic and political forces have put strict bounds around the amount of government-supported scientific activity in the UK. The reasons for this are understandable, even if there is room for debate on just how rigorously they ought to have been applied. This situation is not unique to the UK, and is not a quirk of the political party that happens to be in power. In democratic countries with governments of a different persuasion – for example, France and Australia – much the same trends can be observed. In the United States, where the financial profligacy of the Reagan Administration strongly favoured the science budget for a certain period, many features of the transition are once again noticeable.

The fact that Japan is giving increased support to basic research does suggest that British scientists might be discounting too readily the possibility that the resources devoted to science will in due course resume their expansionary trend. It could be that Britain, along with many other countries, is in a period of review and re-assessment, where the powers that be try to readjust their priorities and their views on scientific and economic pushes and pulls. The indications are that the current state of more or less level funding in UK science will persist for a long period [§1.2]. But this may just be a pause while the economic and social framework is updated and the wealth-creating base adjusts to a much higher level of technology – a level which will require more science and more scientific skills than ever before.

It is not to the point here to discuss how far what has happened to British science in the 1980s and 1990s is due to quite general world-wide tendencies and how far it is due to accidental factors such as adverse economic circumstances, or the deliberate and distinctive policies of the Thatcher and post-Thatcher governments. What cannot be disputed is that the balance of forces within the world of science has changed irreversibly. Instead of being allowed to expand more or less freely under its own steam [§4.1], it is now strictly confined within fixed or very slowly growing budgets. The internal pressures that build up can only be accommodated by a structural transformation involving all sectors of scientific activity, from long-term, curiosity-driven academic research to urgent technological development. This transformation affects the organizations that fund and manage R&D, and the policies they pursue. It affects the education and training of professional scientists, as well as their subsequent careers. In short, it affects all aspects of the scientific enterprise.

It is of no particular significance, therefore, whether or not the total science budget in one country or another is still growing slowly in relation to its GNP. Almost everywhere, the same forces are at work, similar pressures are building up, and a similar structural transformation is taking place. That is our interest here. One of the difficulties, however, is that many of these forces are not internal to the research system. Modern science is not an independent, self-sustaining social enterprise. Historians of science draw attention to the intellectual, personal and institutional connections that there have always been between science and society. One of the characteristics of the recent history of science is the tightening of these connections. Terms such as 'the science base', or 'the research system' refer to complexes of ideas, people and organizations which cannot really be isolated from their governmental, industrial and educational surroundings.

The forces that act both ways between science and society are not simply organizational. Scientific ideas and knowledge pervade the social fabric much more extensively than they did in the past. We are in a science-based, high-technology society.

Influences flow in and out of the research system from and
to almost every other component or dimension of the economy
or the polity – farming, law, inner-city life, tourism, religion,
youth, old age. This is an extremely important point, which
should always be kept in mind. Yet if we tried to allow for it
properly, our analysis of the state of science would soon
proliferate over the whole political and economic scene. To
keep the present discussion within reasonable bounds, we have
to treat this scene as a backdrop in which a great deal is
going on but which will not change in its essential character
over the years.

In particular, we assume quite arbitrarily that the current
financial status of 'science' (in the narrower sense) will persist
for the foreseeable future, in that governments will systematically monitor the funds devoted nationally to academic and
strategic scientific activity and will put very severe constraints
on their further growth. In other words, we assume that science
in the UK – and in most other countries at a similar level of
economic development – will not revert to its historical state
of rapid expansion, but will henceforth have to live within a
bounded envelope of total resources.

For want of a better epithet, we shall therefore refer to this
new regime as science in a 'steady state', as if these bounds
were perfectly rigid. This terminology should not be over-
interpreted. It does not mean that all the changes now going
on are directly or entirely due to current government policies
of 'level funding': it is intended simply to indicate clearly the
distinctive contrast with the 'expansive' regime of the past.

Indeed, this overall financial boundary condition need not
be taken quite literally. Nobody suggests that scientific activity
should not grow at least as fast as the commercial and industrial activity with which it is now so closely linked. In fact,
there are compelling arguments for greatly increased private
sector expenditure on strategic research related to innovative
generic technologies, in order to meet the challenges of international industrial competition [§2.5]. Even where it applies,
as here, primarily to government expenditures on academic
and strategic science, the notion of 'level funding' is reasonably

interpreted to include modest expansion rates comparable with changes in other measures of general economic growth.

Again, for simplicity, we do not exclude the possibility that the scientific community will learn to make a better case for academic research, and persuade the political and commercial communities to provide them with rather more resources than they seem inclined to at present. This also is probably highly desirable, but is most unlikely to be more than a temporary palliative, smoothing the way into the new regime.

It must be emphasized, moreover, that a system in a 'steady state' is never just static. We certainly do not mean that the research system will be frozen into its present form, like some Arctic lake. The use of this term does not imply, for example, that the science vote will be the just the same every year, nor that there will be a fixed pattern of distribution of effort over various sectors of the R&D system. On the contrary, it signifies a highly dynamic situation where continued tension between internal scientific developments and external social demands may produce much more rapid change within the system than was normal in the past. Think rather of a river, turbulent and agitated as it flows swiftly and vigorously between fixed banks.

The behaviour of the quantitative parameters [§4.1] is not as significant in this analysis as what they mean qualitatively. What is important is the changing flux of ideas and people into and out of institutions, and the shifting boundaries between them. The centres of power and the sites of activity are on the move, both sectorially and internationally. The redistribution of effort across the map of knowledge [§2.2] – for example, the decline of high energy physics relative to molecular biology, or the explosion of interest in communication technologies – is also a qualitative phenomenon, amounting in effect to a radical restructuring of the relationships between the disciplines.

Some of this restructuring may actually be due less to the transition to the 'steady state' regime as such than to other historical factors such as changes in the world balance of economic power. The international scientific, technological, economic and political contexts of UK science must be assumed to become increasingly dynamic and challenging [§8.9]. Although a general and complete world-wide transition to 'steady state' science does not seem imminent, many other countries are going through similar changes in the way that they organize their scientific activities. These developments will surely sharpen further the global competition for the rewards of scientific and technological excellence, and make the present analysis even more urgent and significant.

5
Allocation of resources

The King was in his counting house,
Counting out his money.

5.1 The emergence of science policy

In spite of their rugged individualism and celebrated independence of mind, scientists have always sought, and often gained, government patronage. Indeed, in many countries of Continental Europe the academic scientists were mostly, in a technical sense, civil servants, although their 'freedom to teach and to learn' was usually well protected by long-established traditions of operational autonomy for state universities and scientific academies. Paradoxically, in France and Germany, where these traditions are still very much alive, academic science has resisted 'collectivization' more successfully than in countries such as Britain, where academics are not (at least in principle) government employees, and have always been very chary of putting themselves directly in the power of the state.

In the Anglo-Saxon countries, however, it was customary for state support for basic research to be very limited, except in fields such as medicine and agriculture with direct connections to major sectors of governmental responsibility. This custom changed radically in and after the Second World War [§4.2]. Public funds soon became so vital to the advancement of science – especially the academic research carried out in universities and other higher education institutions – that they largely determined its direction and shape.

One might say, however, that until the early 1970s the scientific community largely dictated the terms on which this support was provided. The funding was so generous and so unconditional that it fuelled further expansion of science at a greater rate than the historical average, thus making up for the war-time pause [§4.1]. As long as the scientists could get more or less what they wanted in this way, desultory political initiatives, or analytical exercises in framing a national policy for science, aroused little interest in Whitehall, Washington, Ottawa or Canberra.

The unplanned transition to level funding in all these countries in the mid 1970s [§4.5] caught scientists, civil servants and politicians unawares. It quickly became apparent that some sort of coherent strategy for the overall support of science was needed, if only to provide a rationale for the disappointed expectations, chilling freezes, and sometimes cruel cuts, suffered throughout each national scientific community. Under 'steady state' conditions, science inevitably becomes the subject of 'policy' debate, as the scientists can no longer pull the strings behind the scenes and are forced to contend openly with rival interest groups in the public-expenditure arena [§4.6].

The real significance of the emergence of a concept of science policy at this time is not that it betokens a transition to an orderly and equitable decision-making process with advantageous outcomes for the nation: it simply shows that science had to come in from the cold, into the warmth of the political kitchen. In 1969, Daniel Greenberg provoked a storm when he wrote of the 'politics' of pure science: the French realistically use the same word, politique, for both politics and policy. One might say that the significance of the rise of science politics/policy is that knowledge-making has been 'collectivized' in the political sense – i.e. has come under the control of the grand collective, the nation state. It is a subsidiary question whether or not this control is actually exercised according to a consistent set of conscious principles with attainable objectives.

5.2 Science policy or science politics?

Science policy is very seldom as coherent or decisive as its name suggests. Even in countries practising economic planning, scientific research has proved difficult to incorporate into the general political structure. It is always too uncertain in outcome, too specialized, and too heterogeneous in its applications, to fit neatly into any of the standard bureaucratic pigeon holes. Does space science belong under the military or the civil authorities? Who should manage the invasion of medicine and agriculture by biotechnology? Doesn't anybody know what we should do to discover whether or not 'mad cow disease' is infectious? How did we ever persuade ourselves that we should spend $100 million on this 'cold fusion' nonsense? Responsible bodies do have to make decisions on questions like these, but they seldom evolve logically out of well-informed memoranda and expert advisory reports, referring to long-established principles and assessing soberly the relative costs and benefits of all the options. In other words, science policy is no more 'scientific' in practice than other forms of policy, and is subject to the same historical and cultural influences.

Nevertheless, the leaders of the scientific community, and those with whom they deal in the political world, are forced to take an interest in the systematic analysis of the decisions they negotiate between them. Under pressure to justify particular lines of policy, they cannot afford to be ignorant of concepts such as input/output efficiency, the sophistication factor for instrumentation, or citation counts as scientometric performance indicators. Many of these concepts are highly technical, and are applicable only in the peculiar context of R&D systems, but others are quite general. Laboratory scientists and scientific administrators are thus being introduced to the techniques and insights of the social sciences, which their specialised education in the natural sciences would not have taught them to appreciate.

In the past, if there was such a thing as science policy at all, it was *macro-economic* in its scope. It was about persuading the Treasury to give the Department of Education and Science another hundred million pounds for the research councils, or amending the line item in the Federal Budget allocating x billion dollars for the National Institutes of Health (NIH). The supposition was that more detailed questions, such as whether to finance a new research programme or a new research facility, could be decided on their absolute technical merits, with little reference to other projects that might be competing with them for support. It is notorious, for example, that it was not until the late 1960s that questions began to be asked about the funding of experimental nuclear and high-energy physics, relative to other branches of the natural sciences such as chemistry and biology.

Under 'steady state' conditions, where the demand for financial support for innumerable highly meritorious projects greatly exceeds the supply of funds, the focus of policy has moved to the detailed allocation of resources among vigorously competing programmes or institutions. Decisions that could previously be determined by secret committees, on the confidential advice of expert consultants, become subject to political scrutiny and judicial review, not to mention good old-fashioned pork barrel politics and sophisticated professional lobbying. The normal processes of political life, which previously applied only to really Big Science decisions, such as the UK subscription to CERN [§3.6], or the siting of the US SuperConducting Supercollider, are now being applied to innumerable much smaller projects. The scope of science policy, as a practical art and as a theme for analysis, thus extends much further into the organizational arrangements and interests that influence such decisions.

Discourse on national science policy used to be framed in strictly economic language. The input of funds to something called 'the R&D system' was supposed to produce an output of something called 'scientific and technological knowledge'. This was assumed to be a marketable commodity [§2.7] that was eventually put to profitable use in the creation of tangible wealth. To some extent, this caricature is derived from current political ideologies, which systematically ignore or denigrate public goods, cultural and welfare values, social synergies, intergenerational transfers, and other non-negotiable features of life and work. But it may also stem from the diversity, dispersion and mutual independence of the institutions that traditionally received public funds for research. The internal structure of a national R&D system was obviously much too complicated to be analysed in economic terms, so it might as well be treated as a 'black box' capable of transforming hard cash into somewhat softer knowledge.

What we have learnt in recent years is that science policy, like all policy, has many important aspects that cannot be represented in economic terms. As organizations become more structurally interdependent – for example, within a large but tightly managed multinational corporation – internal budgetary decisions and resource transfers between divisions are determined less by 'economic' factors than by socio-political considerations such as managerial authority, career progression, personal esteem, structural stability, etc. Science is not just being transformed economically: it is also being changed sociologically, psychologically and philosophically. As we shall see, science policy now needs to be framed and analysed in all these aspects.

In the present chapter, however, the economic framework is taken to be primary. We are concerned with the way that the financial stringency characteristic of 'steady state' conditions continually pushes contentious *allocation* decisions to the top of the science policy agenda. How are these decisions made, and what direct effect does this have on scientific activity?

5.3 Scrutiny and accountability

As researchers become more dependent on external funds, the procedures for getting hold of resources loom larger in, and absorb more of, their working lives. Frantic competition for funds means the frantic writing of grant applications, many of which are fruitless. This effect of the transition to level funding is very familiar to everyone nowadays in the scientific world. 'Apply or Die' one might say, mirroring 'Publish or Perish' as a cynical formula for survival.

This snowballing of grant applications is amplified by greater concern by anxiously competing applicants about the way that grants are awarded. This concern is shared by funding bodies, who are under pressure to distribute research resources in larger and larger quantities. Allocation decisions become more crucial, and are therefore made more judiciously, in consideration of a wider range of factors. More work has to be done on the formulation and selection of research projects. Their objectives, methods and budgets have to be set out in greater detail, scrutinized more closely and assessed more formally. The whole procedure must not only be managed on a much greater scale, often with thousands of applications to be processed each year: the processing of each application has also grown much more elaborate and laborious.

It is not just that the choices have become more agonizing – that the satisfaction of the panel with a good day's work in assessing a metre-high pile of project proposals is tempered with the uneasy thought that one of the rejected 40 per cent of 'alphas' might just possibly have led to a marvellous discovery. The real trouble is that choices that could once have been left to the specialized skills of a small group of conscientious experts have to be *justifiable*. There has to be a line of argument that would serve to explain each decision, both to unsuccessful applicants and to higher authorities in funding bodies. In an inherently obscure situation, where intangible clues and instinctive convictions are often the best guides, it seems necessary to pick on a minor point of theory, or a trivial technical defect, as a reason for rejecting an otherwise plausible project. So the application, however grandly speculative in its basic principle and likely outcome, has to be drafted as if it were a legal document, to provide no tiny handles with which it could be thrown out.

The allocation system is thus under enormous pressure to be more *accountable* in its decisions, both to the authorities that supply the funds and to the researchers who apply for them. In practice, these accounts are very very seldom scrutinized, but if a decision is challenged then they have to be available and in good order. This applies with particular force to research funding by government agencies: one of the advantages of charitable foundations as patrons of science is that their officers and/or trustees are not under any such constraints in their distribution of largesse. As was shown by Warren Weaver in his imaginative use of Rockefeller Foundation money to support molecular biology in the 1930s, this freedom can sometimes be exercised very beneficially indeed – but it is only available nowadays in relatively small areas of the scientific scene.

The general change from a *customary* to a *legalistic* framework shows up with particular clarity when questions of deception or misdemeanour are raised. But the whole issue of scientific *fraud* has now become so important in the relationship between the scientific and political worlds that it would require a separate book to describe and analyse. All that we need to note here is that its recent emergence into the public arena is one of many signs of a major transformation in the way that science is now done.

The increasingly elaborate procedures that have evolved in response to these pressures have become a significant element in the overall *cost* of doing science. It is not only expensive in administrative resources to prepare and process a project proposal – to keyboard a text of dozens or even hundreds of pages, print it out, photocopy it, mail multiple copies, register it in a database, recopy and remail to referees, send in reports, collate reports, send documents to panel members, etc., etc. There is also a very substantial hidden cost in the time and efforts of a large number of highly qualified scientific and administrative personnel: the staff of the funding body, the specialist referees, the expert assessors, the senior scientist who sit on the panels and committees, not to mention the researchers who made the proposal in the first place. There are other costs which are even more difficult to assess, such as the opportunities lost through procedural delays, or the talented scientists driven out of research by the excessive paper work it entails. The R&D system has not yet ground to a halt under the burden of these procedures, but there is a general feeling in the scientific world that their supposed benefits are no longer commensurate with their undoubted costs.

It is scarcely surprising, as research resources have become harder to come by, that the administrative machinery for their allocation has grown into a monster that devours a great part of the time and effort that goes into the research cycle. As we shall see, this development is not entirely disadvantageous or wasteful. But it has had the further consequence that the advancement of knowledge no longer depends so completely on the originality and technical skills of the scientists who propose research projects and carry them out if they are approved. People who are not directly engaged in research, such as the senior managers, advisory committees and professional staffs of research councils, government departments, industrial firms, etc. are having to bear more direct responsibility for the science they choose to support. This is an important development which will be taken up later [§5.8].

5.4 From peer review to merit review

The irresistible pressure to make their procedures more accountable has forced funding bodies to be more *systematic* in the way that they assess and select research projects. This means that the selection *criteria* have to be made more explicit. It is no longer sufficient for a research council to say that they are making grants to the 'best' scientists, for research of 'timeliness and promise'. Their advisers have to work through solemn checklists of attributes, indicating whether they think that 'the outcome of the research is likely to make an [outstanding], [very valuable], [valuable], [useful] contribution to knowledge,' or that 'the proposed method is [highly imaginative], [novel, but well conceived], [an extension of standard practice], [well established], [routine]', and so on, as if all such considerations could be weighed up and aggregated, item by item, in coming to a decision.

In practice the actual allocation process seldom works so mechanically, and little attention is given to the ticks and crosses in the questionnaire boxes. But the way these are phrased is a significant indicator of the principles supposedly governing the funding of research [§4.6]. Although they vary greatly in wording from one funding body to another, they define criteria of two types: 'internal' and 'external'. It is important to appreciate the differences between these, and the changing balance between them.

The *internal* criteria are highly specialized, since they relate to the *performance* of the research. That is, they are designed to answer questions about how well the science is likely to be done, which hinge on very technical issues that can only be decided by specialists. Only a real expert can say whether a field of science is ready for further investigation, whether the applicants are scientifically competent, whether the research techniques are sound, whether the apparatus is both necessary and sufficient, etc. In other words, application of these criteria depends fundamentally on 'peer review' – that is, a process relying on the opinions of a small group of persons with close knowledge and experience of research in the field to be investigated.

The *external* criteria are more general in scope, in that they refer to the anticipated *results* of the research: what will be their *scientific, technological* or *social* implications [§2.5]. For example, if the project is designed to test a current scientific theory, is that theory significant outside a very narrow field? Or could the outcome be a step towards an identifiable technological goal, possibly of industrial value? Or might the project help to deal with some familiar social evil, such as disease, crime or poverty? These criteria bring into play much wider issues, where good commercial, political, medical, legal, or even ethical judgement naturally takes precedence over research expertise.

Both types of criteria are obviously important, and both must be satisfied if the research is to be considered worthy of support. But as the allocation process itself comes under scrutiny from the bodies providing the resources, the balance inevitably swings towards the external criteria, which are so much easier to explain to non-scientists. For example, some organizations funding academic science state that they have replaced 'peer review' by 'merit review', indicating that they are taking account of non-specialist opinion on the relevance of the research to socio-economic problems outside science.

The demand for public accountability thus moves the whole research system towards much greater emphasis on ends/means rationality in all its endeavours [§4.2, §4.6]. The way is opened up for quasi-economic talk about the ratios of 'inputs' to 'outputs', of 'costs' to 'benefits', or of 'investments' to 'impacts', not only for science as a whole but in relation to particular projects or research programmes. This concern then extends to the whole cycle of scientific activity, from the conception and formulation of research projects to the dissemination and exploitation of their results.

5.5 Evaluation as a craft

The very reasonable requirement that scientists should be *accountable* for the use of the funds they receive is thus transmuted into a requirement that everything they do should be *evaluated* by or for the donors. The systematic evaluation of the prospects and performance of researchers, research programmes, and research institutions is a typical feature of 'steady state' science. Indeed, it has become so prevalent that it has developed into a veritable craft, spawning specialized firms, international conferences and professional journals to practise, debate and disseminate its methodologies.

In the past, this evaluation, when it occurred at all, was unsystematic, episodic and *ad hoc*. It was not undertaken according to a schedule, but was normally incidental to a particular decision, such as whether to promote a particular scientific employee, whether to award a prize to a particular nominee, whether to continue funding a particular research unit, or whether to support a particular project. Nowadays, every turn in the research cycle may be evaluated, from the conception of a project in outline ('Two sides of A4 will do'), through its formulation as a grant application ('Not more than 50 pages'), successive progress reports ('Explain any deviations from the original research plan'), to a final report of results achieved ('Include copies of all publications arising from your research'). Individuals may be subjected to annual 'Research Performance' or 'Job Appraisal' reviews, and research units, research establishments – even whole universities – may be put through a stringent assessment every few years.

It is very likely that the traditional academic culture was too indulgent of lazy, feeble, ill-conceived or inconclusive research, and that the general quality of scientific work can be considerably improved by systematic, objective evaluation. One must obviously set against this benefit the administrative costs, obstructions, and time delays of any such procedures – especially when these are essentially superficial exercises designed to demonstrate bureaucratic diligence. What is more important, there comes a point where they produce major structural effects, on personal careers, on institutional arrangements, and on the advancement of knowledge itself. Indeed, 'evaluation' has become so intense and pervasive in contemporary science that we shall return later to the discussion of these various effects [§9.4].

What we can say at once, however, is that all genuine scientific activity and achievement is so peculiar to its cognitive context – that is to its particular field of knowledge – that its innate quality cannot be assessed satisfactorily without a large input of highly specialized expert opinion. Stringent peer review of research promise and performance is nothing new in the scientific world. Appointments, promotions, prizes and publication acceptances have always depended on the considered judgements of authoritative, independent referees. The trouble is that to do this properly requires large quantities of the most valuable resource in science: the personal time of the most competent researchers. Such judgements are also necessarily subjective and very local in their scope; it is unwise to give them absolute significance, especially when it comes to making comparisons across a wide range of subjects or institutions. As evaluation procedures multiply and expand across the scientific world, and as the whole business of research becomes more dependent on them, there arises a demand for a cheaper, more systematic and more 'objective' method of assessing research capabilities and prospects.

There are thus very good reasons why *quantitative* methods are being used increasingly to evaluate research. In particular, the computerization of the information system of science makes it relatively easy to extract, sort and manipulate in bulk the bibliographical data contained in scientific publications – the keywords in their titles, the names and institutional affiliations of their authors, the other scientific papers that they cite, and so on. This development has been facilitated by the discovery that the indexing of citations can be made to pay for itself in other ways, as a search and retrieval tool. The mechanical analysis of published papers, abstract journals, citation indexes, patents, etc. not only provides a wealth of systematic information at a tolerable cost: it also generates indicators whose apparent objectivity and comparability from field to field makes them look ideal as formal measures of research performance.

The general arguments favouring the use of *bibliometric* data to evaluate research are very compelling. It obviously makes sense to suggest that a scientist who has published twenty reputable papers in the past ten years has been contributing more to knowledge than one who has published only two, or that a paper that has been cited twenty-five times by other researchers in the past year is of distinctly higher quality than one that was cited only once – and then only by its author. A university department whose staff are extremely productive, in terms of research publications per head, surely deserves a higher rating – and as a consequence, a more generous allocation of resources – than one where this indicator is very much lower. If the notion of scientific 'value' can be quantified at all, then these are the sorts of indicators that we would instinctively seek out to estimate its magnitude.

A closer look, however, quickly reveals their practical limitations. A scientific publication is not an entity with a standard value. Its significance varies enormously, not only in terms of public citability, but from subject to subject. Chemists report their results in numerous short papers; mathematicians publish infrequently but in depth; geologists produce occasional massive accounts of their field work; engineers generate technological advances rather than publications; sociologists write books; literary scholars put much of their research effort into highbrow critical journalism – and so on. Citation practices also vary markedly, and are not always favourable and positive. Learned journals are of very uneven quality in the papers that they accept. Scientific work in non-European languages – even in languages other than English – is usually under-represented in the international data banks. Worst of all, nobody has yet come up with an equitable procedure for dividing the credit for a multi-authored publication among its several – even several dozen – progenitors. The apparently hard figures are nowhere near as objective, unbiassed, metrically equivalent, or arithmetically manipulable as they seem at first sight.

Enthusiasm for the use of quantitative evaluation techniques is thus strongest on the political and administrative boundaries of the scientific world, where these deficiencies (if appreciated at all) are outweighed by considerations of cheapness and simplicity. There is a continuous demand from government circles, for example, for numerical indicators of the 'output' of the research system to which they make such expensive 'inputs'. Indeed, they would like to go further, and obtain quantitative estimates of the 'impact' of R&D on the national economy. Since even the most general input data are, as we have seen, very ill-defined and imprecise, this is a vain endeavour. Nevertheless it is likely to remain a major policy theme in 'steady state' science.

Opinion on this issue among professional scientists is very divided. On the one side, they argue that an objective 'quantity' should carry more scientific weight than a subjective assessment of 'quality': on the other side, they can produce endless anecdotal evidence from their own personal experience showing just how misleading the numbers game can be. Fortunately, this debate does have a practical middle ground, where it is agreed that bibliometric data may well provide information about the quality of scientific work that is as reliable in its way as the views of expert referees. In order to make the most of this complementarity, however, it is essential that these data should be analysed and interpreted in the light of specialized understanding of the personal and social circumstances under which the research has been done and received.

The real danger is that the attempt to express scientific performance and achievement in one-dimensional numerical terms obscures the inherent uncertainty and complexity of the research process. The managerial environment characteristic of 'steady state' science encourages a crude rhetoric of input/output optimization. But management itself is reduced to nonsense if it is always assumed that uncertainty, ambiguity, hesitation, etc. are somehow wet, defensive, self-excusing, ineffective or inefficient.

5.6 Foreseeing the unpredictable

The competitive forces in 'steady state' science come to a focus in the evaluation of research *proposals* – that is, the systematic scrutiny of plans for projects that have not yet been started. This is the moment of truth in the research cycle: until that point, the project is no more than an idea, without solid form; from then on, it is work in progress, or completed. Thoughtful observers of the modern research world recommend a shift of emphasis from this *ex ante* evaluation of proposals to the *ex post* evaluation of achievements, but their recommendations are confounded in practice by the interests and responsibilities that are fixated on the funding decision.

As everybody knows, however, this is the very ingredient in the whole research enterprise that is the most difficult to evaluate. Since the actual outcome of future research is, by definition, still unknown, any procedure for assessing the prospects of research projects and programmes is bound to be unreliable. If the results could be predicted with any degree of certainty, the research would not need to be undertaken: it is as simple as that. This caveat applies with just as much force to formal methods of evaluation as to more subjective procedures. Strictly speaking, they are all exercises in squaring circles.

Nevertheless, pragmatism wins over philosophy. The systematic evaluation of research proposals is an essential feature of contemporary science, if only to *legitimate* funding decisions. However chancy the distribution of research grants may really be, it is politically unrealistic to leave such decisions to sheer chance. There is a case for randomizing the final choice between two highly rated proposals, rather than, say, splitting the available funds between them in a judgement of Solomon – but only after a thorough review has shown them to be equally meritorious.

The essential point is that the formulation of a project proposal and its scrutiny by expert assessors is now a very important part of the research cycle. It is not just a matter of 'grantsmanship' – that is, the artful writing of a plausible prospectus designed to win support for a meretricious piece of work. What we have to appreciate is that the very notion of a 'project' is a relatively novel feature of our present way of doing science.

In spite of the unpredictability of its outcomes, science has always been a highly *intentional* enterprise. The word 'research' conveys strongly this element of premeditation, purpose and perseverance. The problems that scientists try to solve are not mere puzzles, arbitrarily prescribed as tests of skill: they are defined by rational analysis as desirable objectives, often for years of sustained effort. An experiment is not a mere happening; it is a carefully contrived event, planned in advance within a theoretical frame of possibilities. Scientific apparatus is not a mere collection of *objets trouvés* whose meaning will later become apparent through contemplation: it is designed and constructed, often at the very edge of technical capabilities, to generate and/or observe very specific events in very particular ways. Paradoxically, scientific discoveries, the epitome of the unexpected, often arise most remarkably in the most elaborately planned circumstances that the human mind can devise.

The academic tradition does not separate such consciously purposeful actions from other elements of the research process. It is part of the professional skill of every scientist to choose problems, carry out experiments, set up apparatus, etc. in a way that leads to significant discoveries. These particular activities are interwoven with many others, such as speculating, theorizing, observing, analysing, communicating, and arguing, into a smooth and endless braid. It often takes minute historical scholarship to tease them out from the research notebooks, private correspondence and published papers of individual scientists.

The requirement that this braid be transformed into a chain of finite links, each made explicit in advance [§9.3], is thus a radical break with this tradition. In some ways it is not such a bad thing, now and then, for a researcher or research group to have to think out and put down clearly on paper what their objectives are, how they propose to achieve them, and what resources they will need on the way. This can be particularly helpful in team research [§3.7], where collaboration is often hampered by lack of clear understanding of the common purposes of the work. The research itself may also benefit technically from the detailed critique of a project proposal by a panel of independent referees and other experts.

This subordination of personal skill and technical autonomy to group opinion and decision is an inevitable feature of the collectivization of science [§3.9], with profound consequences for institutional arrangements and individual careers. So also is the loss of flexibility in the response that might be made to an unexpected turn in the evolution of the research. If project proposals were taken literally, it would smother the vital spark of *serendipity* in scientific discovery – that is, the freedom to explore seriously and exploit a curious observation or inexplicable phenomenon. These effects on the organization and performance of research are so significant that they come up for discussion at a number of other points in this book.

The issue for the moment is how far it is really feasible to evaluate a project proposal in terms of its expected outcome. Industrial managers know from experience that there is a limit to what can be achieved in this direction, especially in the 'discovery' phases of R&D. A research project is not, after all, strictly comparable with a building project, or the design of a machine, or the statutes of an institution. However vast, elaborate and precise the apparatus for an experiment may be, the place of that experiment on the knowledge map will not become clear until long after it has been carried out. The 'anticipated profits' section in its prospectus can only be filled by conjecture – not too unrealistically, but speculative nevertheless.

In practice, the most reliable predictor of a valuable outcome of a research proposal has always been the track record of the proposer. As every punter knows, 'form' is the best guide to success: put your money on a previous winner and you have a more than even chance of a good return. This has a very obvious and familiar effect. Since research success depends to some extent on having adequate resources to start with, the allocation of further resources on this basis is a sure recipe for a process of 'cumulative advantage'. Not surprisingly, it was in the world of science that Robert Merton originally observed the operations of the Matthew Effect [§3.9]. This applies nowadays to the selective funding of research projects [§9.5] as well as to the age-old competition for the intangible insignia of scientific merit.

Unfortunately for the cynical sociological observer, this phenomenon has a sound basis. Scientists do vary enormously in research talent, and especially in a capacity for identifying significant scientific problems and thinking out ways of making progress in resolving them. But this talent is not necessarily synonymous with the sort of public reputation that wins prizes or prestigious posts, and does not necessarily endure through a long career. This is the basic reason why rigorous and systematic evaluation, using quantitative and qualitative methods, of the past work of individual researchers, research groups, and research institutions is now becoming such a salient feature of 'steady state' science [§9.4]. It also explains the widespread concern about how to identify this talent early on, to avoid the Catch 22 situation where young researchers do not receive sufficient resources to build up the record of achievement required to earn them.

To what extent can the *strategy* of a research project be evaluated, independently of the talents of the researchers and the technical excellence of the proposed methodology? Is there any way, apart from sounding out the opinions of the best informed scientific experts, of deciding whether the proposal is well founded in principle, and has good prospects of producing a significant contribution to knowledge? Enthusiasts for quantitative methods have developed *scientometric* techniques for mechanically surveying and detecting patterns in the evolution of the research or patent literature. Sophisticated software is used to map out the co-citation, co-author or co-word linkages between published papers, in search of previously unrecognized connections or clusters that might be signs of new growth areas of science or technology [§2.2]. It has to be said, however, that the results obtained by these methods are even more suspect than the quantitative indicators widely used for evaluating research performance, and are only useful as a way of reminding expert assessors of significant new developments that might otherwise be overlooked.

For all its weaknesses, peer review of research proposals is thus bound to remain a key procedure in the detailed allocation of resources in 'steady state' science. Yet agreement on how best to carry out this procedure seems unlikely, and its susceptibility to human frailties continues to be a source of much uneasiness, within and outside the scientific world. Different agencies, in the UK and elsewhere, have developed a wide variety of administrative procedures for arriving at well-founded, unbiassed opinions on research proposals. It is obviously necessary, for example, to counter the natural tendency of a narrowly specialized panel of expert practitioners to serve their own cognitive and material interests, both by the inclusion of non-experts and by other administrative means.

The sheer cost of all peer review procedures, especially in terms of the time demands on highly skilled researchers, is also an important consideration. The notion that greater effort would make them more dependable is confounded by the inherent uncertainty of research activity. Turning up the gain control does not make a noisy information channel more intelligible. It is very likely that they have already been multiplied and elaborated beyond the point of economic efficiency in research funding, and should be applied at longer intervals to larger items.

The main source of uneasiness lies deeper. It is a commonplace of the history of science that the conventional wisdom has never been a satisfactory guide to scientific advance over the longer term. It is true that there is no convincing evidence that reliance on peer review is a recipe for a programme of unadventurous projects with all too predictable outcomes. Nevertheless many scientists fear that if all research came under its sway it could inhibit genuine, if apparently wildly eccentric, originality. The fact is, perhaps, that increasingly detailed scrutiny of research proposals is bound in the end to yield diminishing returns and may, by putting too much emphasis on their foreseeable aspects, ultimately prove counter-productive.

5.7 Balancing priorities

Present-day political and economic doctrines teach that resources should be allocated according to pre-determined *priorities* related to their expected benefits, in the shorter or longer term, to national prosperity, security and welfare. It is not clear historically whether these doctrines have deep roots in principle – for example, in the perennial call for the planned application of science to social needs – or whether they are simply a pragmatic response to the increasing demands of an ever-growing social activity. All we can say is that 'priority-setting' is a major feature of 'steady state' science.

But how should this be done? It is all very well to appoint a prestigious 'advisory council' of scientific and technological worthies to report to the head of government on progress and prospects in various areas – but who will ensure that its recommendations are put into effect? Contributions from Parliamentary Select Committees may not always be welcome, and are usually ignored. Perhaps there should be a strong, single, Minister for Science, preferably with a seat in the Cabinet, supported by a team of civil servants to work out a schedule for the coming decade – and with the power to impose it.

Such procedural issues are the staple of public debates on science policy in every country [§8.9]. They are manifestations of the general process of 'collectivization' at the top political level. But they are also very complicated and diverse, because they are really much more about government in general than about science [§5.2]. Being so near to the centres of political power, the science and technology policy machinery has to conform with the mechanisms used for setting priorities and resolving conflicts across the whole business of government in each country. The way they arrange such matters in Washington, or Paris, or

Canberra, or Beijing, may be very interesting, but throws little light on how it should be done in Whitehall and Westminster.

The well-meant attempt by UNESCO, a few years ago, to lay down basic objectives and organizational models for 'national R&D systems' was thus largely fruitless. On the one hand, the formal objectives of any such structure are always trite, and wildly over-optimistic in relation to national realities: on the other hand, the appropriate organizational arrangements will always depend on the national political culture and historical situation. For example, take half a dozen countries with a federal constitution – Australia, Brazil, Canada, Germany, Nigeria, and the United States – and ask yourself whether you would advise them all to use the same formula for the distribution of their science policy responsibilities between their federal and state governments. All that such exercises show is that this is now a matter of universal concern, from the largest and most highly industrialized nations to small countries that are still largely undeveloped economically.

Indeed, talk of a 'national R&D system' is a political euphemism, suggesting a more orderly structure than the reality. The institutions that undertake scientific, economic and social, humanistic, technological, medical, and agricultural research are usually as heterogeneous as the bodies that deliver educational, legal, commercial, industrial, medical, military and social services to the community at large. In the UK, for example, they would include universities, polytechnics and colleges, government research establishments, research council units, contract and consultancy firms, regulatory bodies, charitable foundations, professional institutions, learned societies, industrial laboratories, hospitals, and field stations, spread across all sectors of the economy and polity.

Nevertheless, these very diverse organizations do form a 'system', in that they are interconnected in many dimensions. Researchers in a hospital, a university medical school, a medical research council unit and the laboratory of a medical charity laboratory may be working on the same approach to a treatment for cancer. Techniques may be developed in a Ministry of Defence Research Establishment in close collaboration with scientists and engineers in an industrial firm. A pharmaceutical company may fund a specialized research unit in a university chemistry department. An environmental protection body may be chaired by an eminent academic scientist, and may be in regular contact with the institutions where their staff were trained.

The elements of this system thus habitually interact with one another in their day-to-day affairs, frequently coordinate their separate plans and policies, and often have to rely upon one another without contractual obligations. Indeed, they make up such a closely woven web that it is not always obvious how the various strands should be differentiated and grouped together. This is one of the reasons why the statistical indicators for R&D expenditure are so uncertain and difficult to compare from country to country [§4.5]. A particular type of military R&D, for example, might be carried out in a university in the United States, in a government laboratory in France, in an industrial firm in the UK – and supposedly not at all in Japan. It is not simply a matter of disentangling such items from the bulk data: the categories under which they might then be re-assembled would depend on the precise purpose of the comparisons to be made.

Even when a schedule of priorities has been drawn up, its implementation is usually frustrated by the difficulty of finding out how much is actually being spent under various headings, and for what precise purpose it is being spent in each case. At the national level, R&D funding is not only a statistical hotchpotch: it is also a budgetary rag-bag [§4.1]. If the overall allocation of resources is to be governed by national priorities, then comparative information is needed about the objectives and resources of a great many quite disparate research programmes, institutions, sectors of the economy, national interests, etc. Some of the component inputs and outputs of research can be roughly estimated in money terms but the considerations that initiated them are so varied, and many of the factors affecting their results are so uncertain in value or outcome, that formal cost–benefit analysis is barely feasible, if not quite misleading.

In a pluralistic society, the overall allocation of resources is complicated by the fact that individual industrial firms and government departments undertake R&D in support of their regular operations, according to their own internal priorities. An environmental protection authority, for example, may have to mount a strong basic research programme in ecology simply to meet its statutory requirements. Much of the chemical research supported by a pharmaceutical company may be shaped by the need to protect its intellectual property rights. It is difficult to bring more general policy criteria to bear on such priorities, even though the science involved may have important consequences outside the firm or department.

Indeed, research funding becomes such an object of desire for bodies such as universities and research establishments that they do not scruple to use all the influence they can muster to win it. It becomes very obvious, for example, to a university that is not highly ranked for research quality, that even its best prepared projects are likely to get a 'low' priority – which in this game means no priority at all. Instead of lying back and accepting meekly the verdict of the experts who advise the funding bodies, the university is strongly tempted to try to obtain what it wants through other channels. In the United States, the whole priority-setting process at federal level is being undermined by the increasing proportion of 'pork barrel' funding of mediocre research facilities and programmes, voted through by legislators keen to demonstrate their partiality for the folks at home. This phenomenon, although perfectly legal, is very embarrassing for the leaders of the scientific community, because it takes the allocation of research resources out of their capable hands and leaves it to the much less rational outcome of very diverse and arbitrary political forces. Nevertheless, this is a very natural development under 'steady state' conditions.

Another major complication is that it can no longer be assumed that every form of R&D activity can be assigned a slot along a spectrum of 'relevance', from the most basic to the most applied. Each successive segment of this spectrum could then be supposed to be the main concern of one of the institutional sectors of the R&D system, from the universities at one end to private industry at the other. Although this 'linear model' has always been a gross over-simplification of the sources of knowledge and invention, there was a time when the total R&D expenditure in each sector could give a rough idea of the way in which national

R&D resources were distributed in terms of relevance and imme-diacy of outcome. Thus, a UK decision to reduce the research component in university block grants could have been defended as a straightforward way of following the Japanese policy of giving a lower priority to fundamental research than to its applications.

But this model is now entirely misleading [§2.5], in that it suggests that there is still a sharp distinction between the 'aca-demic' type of science undertaken in universities, and the 'applied' science typical of industrial R&D. The present situation is not nearly so simple or clear cut as it used to be. One of the characteristics of 'steady state' science is that institutions become involved in programmes covering a wide range of urgency and relevance. University departments and research council estab-lishments have to earn what they can in commissioned research, while industrial firms must maintain competitive capabilities in basic science to keep informed of new developments.

Thus, the way in which R&D activity is distributed across this spectrum no longer mirrors the relative flow of resources into different sectors or institutions. The total expenditure in the UK on 'curiosity-oriented' basic science, for example, is not a line item in any single budget, but has to be estimated from the detailed analysis of the budgets of a large number of different organizations. At the same time, to return to the Japanese case, *their* effort in basic science would be seriously underestimated if one neglected the amount of this that they were doing in *their* industrial firms.

This is more than just a complicating factor in the compilation of an *Annual Review of Government Expenditure on Research and Development*, or a volume of *National Science Indicators*. It has serious policy repercussions. It is much more difficult now than it used to be to decide where to place a particular R&D project on the traditional spectrum of 'applicability'. Nobody doubts that decisions about the relative distribution of resources between the two ends of this spectrum have still to be guided mainly by convention, experience and imagination rather than by formal economic analysis. But between these extremes [§2.5] lies much research that is essentially 'basic' – i.e. not obviously applicable – and yet not altogether unlikely to produce 'exploitable' results.

The question whether it is directed towards a high-priority objective can only be answered by a highly subjective assessment of its 'strategic' relevance to that objective. It is also bound to depend on the context of that assessment. Thus, the university scientist who is personally motivated by sheer curiosity to undertake research on, say, the theory of superconductivity, might privately describe the prospects of his project in rather different terms from those that persuaded a committee of the Science and Engineering Research Council (SERC) to award him a research grant, in the hope that it would eventually contribute something to electrical engineering.

It is typical of the present situation in science that almost every scientific worthy who gets up to speak in public about the funding of research starts off with an attempt to define, or redefine, distinct categories of R&D along the traditional spectrum. The obvious inadequacy of all such schemes is one of the signs that science and technology have moved into a new regime where distinctions between research projects in terms of their manifest 'relevance' are no longer meaningful. The trouble is that the scientific community has failed to develop more subtle measures of the contribution of R&D to the political, economic and social needs of the nation. As a result, politicians and civil servants who know little about science impose their own value schemes on it, and evaluate research activity according to this naive utilitarian yardstick.

There can be no objection in principle to setting general priorities among the various long-term objectives of a national research system, and allocating resources accordingly. It is quite proper to decide, for example, that agricultural research should now be directed more towards environmental protection than towards gross productivity, or that strategic research on new materials should take precedence over support for the development of conventional engineering products. Whether or not such broad-brush decisions can ever be sufficiently well-informed to have the intended effects, they arise naturally out of the normal responsibilities of a democratic government towards its taxpaying citizens.

But, even at this level, the demand for accountability requires these priorities to be formulated in a strictly utilitarian language which inevitably affects the overall distribution of funds between the various sciences and their associated technologies. Thus, for example, when the UK 'Science Research Council' was renamed the 'Science and Engineering Research Council', this was the political signal for a reorientation towards research that seemed more clearly related to national economic and industrial needs. Similarly, when the 'Social Science Research Council', was renamed the 'Economic and Social Research Council', this indicated that it should no longer view itself as the foster mother of various academic disciplines but should take the initiative in research on topics of direct societal significance, such as health economics, industrial innovation and drug addiction.

The general effect of collectivization [§3.9] is thus to move resources away from apparently 'useless', curiosity-driven, fundamental research towards problem areas and disciplines that can promise answers, in the relatively short run, to currently perceived economic, political or social questions. Funds are set aside for programmes with specific practical purposes, such as combating AIDS, investigating acid rain, or assessing the performance of school children, and handed out to research groups in a number of different institutions in the private and public sectors. Efforts are also made to identify other areas where potentially exploitable results [§2.3] are emerging, and to give them special support. The scientist devoted to the traditional enterprise of 'pushing back the frontiers of knowledge' in 'the honest search for truth' simply does not figure in the priority scheme [§9.7].

These overtly utilitarian priorities have another, less obvious, effect, typical of 'steady state' science. Pushed down politically from the top, and pulled down by fiercely competitive researchers from the bottom, they become influential at the lower levels of the funding machinery and lend unwarranted weight to 'external' criteria in the assessment of individual grant applications and other detailed proposals. Projects that ought to be evaluated strictly on their technical and scientific merits are given marks for their apparent relevance to particular economic, political or social problems [§5.4]. An inexpert 'merit' reviewer is swayed by an unsubstantiated hint that a pedestrian piece of pharmacological research might lead soon to a cure for cancer – or, reflexively, an honest academic physicist is tempted to emphasize the very doubtful practical implications of his work on superconductivity rather than its novel theoretical power [§9.9]. The point is that priorities of this kind should have already been taken into consideration in the distribution of funds over broader areas of science and technology, and should not be allowed to play a double role in the allocation process.

The attempt to take general command of the R&D agenda in the name of national priorities is a recent development whose outcome is not yet assured. The pluralistic political tradition is that this agenda evolves through a multitude of interactions in parallel, but loosely linked intellectual and commercial, marketplaces. It is highly questionable whether any single agency can acquire sufficient grasp of the detailed results of research – and of the needs it might serve – to assign realistic priorities for the allocation of further resources [§9.8]. Everybody now appreciates the practical impossibility of planning in advance, from a single centre, the routine manufacture of all manner of standard products to meet the foreseeable needs of a nation: it is scarcely credible that this approach could succeed where every item is novel, where the means of production are uncertain, and where the needs to be met are not even clearly conceived. Nevertheless, it is characteristic of 'steady state' science that conscious efforts are made to map what is known and what is needed, and to match the knowledge against the need.

5.8 From the allocation to the management of resources

Although they would not want to admit it, the science and technology policymakers in an operational government department such as the Ministry of Defence have a comparatively easy job. Having established the objectives, relative priorities and budgets for their various major R&D programmes, they can allocate resources in detail through responsible officials down the line, and get to work. Apart from a small amount of open-ended fundamental research with distant 'strategic' horizons, this will consist principally of research projects with clearly defined applications, together with a great deal of technological development, design, and demonstration of the kind required, for example, to bring a new weapon into production and use. Indeed, the boundaries between these later stages in the innovation cycle are so fluid that it is a matter of practical management to decide the point at which they should cease to be treated as continuations of the original R&D effort, and start to be considered the initial steps in the processes of manufacture and marketing under commercial auspices. In other words, whether these activities came under the public or private sector, they have always been organized according to an 'industrial' model [§4.2], and have not been affected structurally by the transition to 'steady state' conditions.

The situation is very different for a body such as the Agricultural and Food Research Council, whose principal responsibility is long-term 'strategic' research over a wide area of national life. The relative priorities attached to broad fields of activity, such as 'animal husbandry' or 'crop protection', can be announced in the Corporate Plan, and reflected in the sums appearing under the corresponding headings in the Council's budget. But these priorities cannot be applied directly to basic research projects in cellular biology, say, which might contribute to practical progress in any one of a number of fields [§2.5]. Nor can it be assumed that a sufficient supply of scientifically meritorious projects will be proposed by academic researchers to take up the resources allocated to each high-priority field. To give a medical example: basic research related to bedsores does not, apparently, interest biomedical researchers, despite the clinical and economic seriousness of this affliction.

The political necessity for such bodies to show practical results not only drives them towards more directly utilitarian projects, involving closer association with 'users' such as farmers, doctors, engineering firms, utility supply corporations, etc.: it also encourages them to adopt a much more *pro-active* stance in their allocation procedures [§8.8]. Economic and social priorities not only govern the choice of proposals submitted by researchers: they also generate 'initiatives' on topics which the researchers themselves might not have thought of proposing. Thus, for example, the explosive expansion of information and communication technology (ICT) in the 1980s indicated the urgent need for fundamental economic and social research on the impact of this development on everyday life. Although it was recognized that there were many academic social scientists with the conceptual interests and methodological skills to undertake such research, they were being very slow to recognize that ICT provided a range of topics worthy of their efforts. For this reason, a substantial sum was earmarked in the budget of the Economic and Social Research Council for a major research programme in this field.

The actual procedures that have evolved to start and run such a programme are very varied, and do not conform to generally agreed principles. They include commissioning an expert to 'coordinate' the whole initiative, convening a 'task force' or 'workshop' of potential researchers and/or 'users' to develop a coherent set of themes and issues, inviting outline project proposals from selected research groups, formulating and advertising an open 'request for proposals', reviewing and selecting projects for both their technical merits and their relevance to the subject, negotiating variations in the project proposals to avoid overlaps or to encourage collaboration, arranging periodic meetings of the various research teams to coordinate work in progress, and encouraging the dissemination of the results of the research when it is all over.

What is happening, in effect, is that research councils and other funding bodies are moving from resource *allocation* towards resource *management*. The 'programme advisory committee' or 'directorate' that is responsible for an initiative may have very limited formal authority over the dozen or so independent research groups to whom they distribute their funds, but informally they find themselves saddled with delicate managerial responsibilities. A creative 'working party' that includes fiercely competing research specialists has first to be coaxed into a cooperative, openly communicative, frame of mind [§9.9]. But, when it comes to project selection, there must be no suspicion of favouritism between the applicants. Negotiations to improve the balance of the programme, decisions on fortuitous operational changes, pressures to ensure exchanges of intermediate research findings, suggestions for active collaboration between complementary research teams, schemes for integrating, presenting and disseminating the results of the programme – all have foreseeable effects on the careers of researchers, research groups and research institutions. Further strains are created by the very reasonable insistence of higher financial authorities that the performance of the research and its practical impact should be evaluated [§5.5]. The future of a whole research community may thus be shaped by the way in which such an initiative is conducted [§9.8].

A research council funding university research in a 'responsive mode' – that is, simply handing out grants to a selection of unsolicited project proposals – can avoid all such responsibilities. Indeed, it may emphasize an academic ideology of strict impartiality and neutrality in its decisions, protecting the anonymity of its peer reviewers and forbidding direct negotiations between its clients and its advisory panels. If there is any conscious shaping of the research community or building up of research capabilities in a specialized field, it is done *sub rosa*, outside the formal agenda of the grants board or committee. The question arises whether the light-weight administrative machinery typical of such procedures can satisfy the heavy expectations of those who now rely for their livelihoods on these decisions and whether funding bodies themselves have the managerial capacity to carry out these new roles.

The increasing tightness of the relationships between funding bodies and researchers is characteristic of 'steady state' science. As allocation decisions become more crucial, their personal and institutional implications become more salient for all concerned. Researchers complain about the additional administrative work that they have to do to satisfy the requirements of financial accountability [§5.3]; they also object to the way that a 'directed' research programme may distort what they regard as the natural development of their subject. Physicists may feel, for example, that a programme of research on the mechanical properties of superconducting alloys, although of obvious practical importance, is unlikely to answer the fundamental scientific question of why they are superconducting. A sociologist may feel that the temptation to enter the competition for a grant from a generously funded programme of research on a current issue of science policy may divert her from her long-term career goal of developing a general theory of the relationships between knowledge and power. Some of these objections are probably misconceived, in that they arise from the prejudices against interdisciplinary research [§3.8], or fail to recognize the challenge of real-world problems to basic theoretical understanding. Nevertheless, they are the outward signs of the stresses induced on academic science by an environment where the emphasis is continually on potential exploitability [§9.7], the promise of utility and the expectation of material gain.

5.9 Command systems and market models

The allocation of resources reaches down, eventually, to the laboratory, library, field site or writing desk, where the research is actually done. As we have seen, scientific work nowadays is seldom a solitary activity. Even when a particular piece of research is written up as if it were carried out by just one researcher acting entirely on her own, this will usually have been in a working environment involving systematic collaboration with a number of other scientists with similar or complementary interests and technical skills.

Most research nowadays is the collective product of what one might call a research entity – that is, a 'group', 'unit' or 'centre' comprising researchers, students, technical staff, etc., working closely together in a well-defined scientific specialty or problem area [§3.7]. This may range from a large, permanent, closely knit, interdisciplinary team under a strong leader, down to a voluntary association of a few researchers who happen to find it convenient to cooperate loosely for a while. Indeed, in discussing academic science, it is often convenient to count every archetypal pure mathematician or ancient historian as a distinct research entity, even though this entity may only have one member.

In most contemporary scientific institutions, it is the research entities, rather than the academic 'faculties' or discipline-based 'departments', that are the basic organizational units for research. They are the originators and performers of research projects, and the active agents for scientific change. The resources for research must eventually come into their hands, either as a budget allocation from within their own organization or institution, or as a project grant from an external funding body. The success of any 'research system' depends on whether these resources flow in sufficient amounts, without unnecessary delay, to the most appropriate entities for the desired activities.

An obvious way of ensuring that these channels are kept in good repair would be to extend the priority-setting/budgetary-allocation mechanism itself right down to the bottom of the system. In effect, funding bodies would then become self-contained research organizations, supporting and running their own research entities. This would be entirely consonant with the principles of 'R&D management', which have long been the norm in industry, in many government research establishments, and even, in some countries, within academia itself [§4.2]. As we have seen [§5.8] there is a natural drift of funding bodies towards more direct managerial control of the research work they support. For all their bureaucratic tendencies, such 'command structures' are usually necessary, and often highly effective, in applying multifarious scientific and technological capabilities to perceived problems. We have only to think of the achievements of the pharmaceutical industry to see what can be done in this way.

But the travails of the Soviet Academy of Science are a dreadful warning against trying to organize basic research along those lines. The superficial mechanical efficiency of a command structure is continually degraded by internal politicking, spurious quality and performance indicators, structural rigidities and other familiar organizational maladies. In the absence of clearly definable goals and objective measures of achievement [§5.5], the allocation of resources to research entities either follows a rigid formula, or becomes an occasion for the exercise of unmerited political influence.

Paradoxically – and perhaps providentially – the transition to 'steady state' conditions has not entirely 'industrialized' or 'bureaucratized' academic science [§6.6]. In the UK, the USA, and other countries with similar university systems, it was essential for state-controlled funding bodies to respect the traditions of institutional autonomy and academic freedom, even as science was becoming almost totally dependent on the state financially. This was achieved through several decades of prosperity by the expansion and institutionalization of 'peer-reviewed project proposal' procedures, which were then expected to work in quite a different economic climate. In fact, these procedures have adapted remarkably successfully to this change, and have even been applied within, or on the boundaries of, organizations such as government research establishments which had previously been run on hierarchical bureaucratic lines. The current ideological trend in favour of 'market' systems is only partly responsible for this development, which has its own tradition and rationale in the world of science.

5.10 Research entities as small firms

The tightening of the financial screw on the science base has had a twofold effect. It not only encourages the sub-division of academic institutions into relatively self-contained research entities: it also forces these entities to put a great deal of effort into obtaining sufficient resources to keep alive. As we have seen, only a small proportion of the labour that goes into applying for funds produces any return [§5.4]. Project proposals not only have to be technically sound: they must also be closely tailored – at least in appearance – to the objectives of funding agencies. They have to conform as closely as possible to current general conceptions of what constitutes promising research, even though they really stem from the idiosyncratic notions and enigmatic hunches of individual researchers [§9.9] .

The result is that many research entities, even within universities, are run now as if they were specialized research *contractors*, vying with one another for commissions to investigate questions that happen to be favoured by highly discriminating customers. Budgets that have been subdivided and allocated downwards through a command structure are distributed to the researchers themselves by something very close to a market mechanism. In effect, the funding bodies have moved on from the patronage of research [§4.2] to its *purchase*. Each of their panels is allocated a block of funds to buy the best research it can in a particular field, and thus generates around itself a highly specialized market of research entities offering to supply the particular products for which there is this demand.

In countries such as France – and formerly in the 'socialist bloc' – where much of the science base is still run on civil service lines, the 'market model' is no more than a metaphor for the informal competition for power and resources within the organization. But where, as in the UK and the USA, there is much greater dependence on peer and merit review procedures, the economic analogy is acquiring more and more reality. The language of 'grants' and 'projects' is giving way to a more legalistic terminology of 'bids' and 'contracts', indicating a more specific accountancy of what is to be delivered in return for what resources.

The precise terms on which a research entity actually obtains its resources are clearly of the highest importance. Most academic or quasi-academic research entities are quite small – typically, half a dozen fully qualified researchers, with another dozen or so students, research assistants, and support staff – and are thus susceptible to the economic hazards of all small businesses. With a very lumpy income from a small range of customers, and no financial reserves to fall back on during a run of bad luck, they find it very difficult to maintain a steady cash flow. Until recently they have not usually been held accountable for costs that would otherwise be designated as 'overheads' – premises, administrative services, infrastructural facilities such as computer access, capital goods in the form of apparatus, etc. – nor for the salaries of tenured academic staff [§6.7]. But in the harsh economic climate of 'steady state' science, the tradition of unconditional institutional support for such items has been seriously eroded, and there are seldom sufficient resources to guarantee continued employment for other research staff [§7.11].

The fragmentation of research funding into large numbers of relatively small, relatively short, projects is a natural consequence of 'steady state' pressures on the allocation system. In some political circles it is axiomatic that financial stringency fosters more efficient use of the available resources. But this is a general economic argument that needs to be justified specifically wherever it is applied. The vulnerability of research entities to short-term financial fluctuations affects the research system in a variety of ways. In later chapters, we shall be looking at the indirect effects of this on institutions and individuals. The question here is how this continuous scrabble for resources influences the *performance* of research, in terms of orientation and quality.

5.11 Competition in project markets

The supreme virtue of a market system is that it is supposed to optimize the allocation of resources through the 'invisible hand' of *competition*. In any particular situation, however, this grand socio-economic generalization requires detailed justification and qualification. Before one can work out what actually is being 'optimized', questions have to be asked about the balance of forces between customers and vendors, the stability of the medium of exchange, the availability of information about the quality of the goods, and so on. The intense competition for resources that is characteristic of 'steady state' science was an unintended – often deeply deplored – consequence of the transition to level funding. The market situation thus generated is very far from 'perfect' in the classical economic sense, and does not necessarily operate to produce ideal results from a scientific or social point of view [§9.9].

What we have, of course, is not a single great market for 'science', but a large number of nearly independent markets for research projects in a great variety of specialties. Although the priority-setting, budgetary processes at higher levels in the funding system may involve direct comparisons and notional trade-offs between major fields of R&D, these are usually more 'political' than 'economic' in form [§5.7]. There is seldom any occasion for direct competition between, say, a project for studying the immunological response to a constituent of an HIV strain and, say, a proposal to search for evidence of planetary systems around nearby stars. Indeed, it is very likely that the latter project would not even have to compete directly with, say, a proposal for observing the surface of the distant planet Pluto, which would probably be directed to a different panel of the funding body.

The most serious feature of these innumerable specialized project markets, however, is that they easily become monopsonies. In a supposedly centralized governmental structure such as the UK, most of the 'customers' for academic research – that is, the research councils – are systematically differentiated in their interests. They even make administrative arrangements so as to avoid the supposedly wasteful situation where two or more councils, committees, boards, or panels are funding almost identical projects. In other words, they are enjoined never to compete directly for the same commodity. This is probably unavoidable in a system where all the funds eventually come through the same channel from the same pocket in the public purse. Unfortunately, there is also a tendency for charitable bodies to define specialized niches for themselves in the type of patronage they offer.

The result is well known. What a typical research entity has to offer is an expertly conceived project in a highly specialized scientific field. It is the responsibility of, say, a research council official to direct the proposal for such a project to the 'appropriate' committee, in his own or some other organization. There it will be evaluated and selected in direct competition with all other proposals of a similar kind that are currently on offer. In effect, for the time being this committee may be almost the sole customer in this particular market, and is not under any competitive pressure from other customers to optimize its choices.

Some of the imperfections of monopsony in project markets can be reduced administratively by making sure that customer groups such as research grant panels are widely representative of the potential 'users' of the research results, that they are not dominated by an oligarchy of specialist vendors, and that all proposals are subject to independent peer review [§5.4]. But the only 'market' solution to this problem is to facilitate direct competition between customers in the same specialized market. This is the situation in the United States, where less centralized arrangements for funding a much larger volume of academic science make it worth while for research entities to submit essentially the same proposals simultaneously to several independent funding bodies.

The conventions of 'responsive mode' project funding do at least entitle research entities to have their proposals assessed on their intrinsic merits, and put the onus directly on customer committees to select wisely among them. The trend towards 'directed mode' funding [§5.8] could have much more serious consequences for specialized project markets. In effect, this means that the customers call for *tenders* to provide research services to meet stated specifications. But, as we have seen, the most likely participants in the tendering process are precisely the research specialists who need to be consulted – often as an invited group – in drawing up these specifications, and who may thus gain 'insider' advantages in the subsequent competition. In the end, this leads to a situation where research contractors become mere clients of the organizations which commission research projects from them: a decentralized 'command' system [§5.9].

What strong competition can really optimize, even in a specialized monopsonic market, is *quality*. There can be little doubt that the pressures of 'steady state' funding have forced researchers and research entities to put a great deal more thought into the planning of projects, in the sure knowledge that these will be subjected to very close expert scrutiny. Scientists nowadays often bewail the fact that only a fraction of all the 'alpha-rated' proposals that are submitted to funding bodies are successful: this could be interpreted as evidence that the general quality of the projects now being proposed has risen to a much higher level, and that only the best of these are actually going ahead. This argument must be taken with a pinch of salt, since it assumes that research is optimally conducted as a sequence of pre-planned projects, and that these projects can be reliably assessed and selected to give optimal outcomes. But, as we have seen, these assumptions are so deeply embedded in the rationale of publicly funded science that they can scarcely be questioned in this context.

Does this mean that market competition also exercises quality control over the 'vendors' of research – that is, over individual researchers and research entities? This is a much more complex and subtle issue, since it involves larger and longer-term personal and social factors. The public reputation of the applicants plays an important, but often informal and unacknowledged, part in the assessment of projects. On the one hand, it is not easy for a scientist without a track record to enter the market: on the other hand, the 'Matthew Effect' [§3.9], exaggerates the principle that the scientists leading the research entities that win the most grants must be the best, and should therefore be given larger and larger resources. Funding bodies, as shrewd customers in a buyer's market, are aware of this distortion, and try to get better information on the intrinsic value of the goods they are purchasing. Instead of relying on market forces alone to control quality, they insist on systematic evaluation of the outcomes of research and of the actual performance of research entities [§5.5].

There is certainly much to be said for making the reputational status of scientists and scientific institutions more explicit and transparent, right across the board. Information on the past performance of research entities is an extremely important factor in project markets, and has significant managerial implications within universities and other research organizations. It is an axiom of economic theory that reliable information makes market competition more efficient. The expert judgement of peers is not necessarily unreliable, but it needs to be backed up with information on the actual contributions of researchers to the literature of their specialties.

The classical market model is too static to apply to a dynamic social institution such as science. The notion of 'resource allocation' does not make sufficient allowance for the *structural* effects of competition when these resources are limited. All that is really 'steady' about the system is its overall envelope of funds. Within this boundary, it is in a state of turmoil and turbulence. Research entities grow apace, or die unlamented. Collaborative alliances and mergers are engineered, to share facilities and research programmes. Radical changes occur in the managerial, financial and legal status of research entities within their parent institutions. The terms of employment of all personnel, from the world-famous research professor to the laboratory technician, come into question. The traditional symbiosis of teaching and research is disrupted [§7.2]. The concentration of research capabilities into a few specialized centres of excellence [§6.8] leaves inadequate support for other fields. Resource allocation goes international, to attack global issues or to meet priorities that cannot be satisfied within national frontiers [§8.8]. These are some of the *dynamic* processes that are typical of the new regime.

6
Institutional responses to change

To market, to market, to buy a fat pig!

6.1 Research institutions

Research is not undertaken by individual scientists or by specialized research groups; strictly speaking, it is undertaken by *institutions*. The basic organizational 'entities' that typically devise and perform research projects are usually too small to stand alone as independent enterprises [§5.10]. It is true that there are now many commercial firms, ranging in size from self-employed individuals to sophisticated outfits employing dozens of scientists, engineers and management experts, offering research and consultancy services or even undertaking speculative near-market R&D in the hope of producing saleable technological innovations. But an enterprise trying to live on grants to do basic science cannot be financially viable unless the overheads and uncertainties can be spread over a large number of different research projects. Although research entities may behave very much like independent small firms in bidding for research projects, they are almost always embedded legally and organizationally in what we might define generally as *research institutions* – that is, substantial corporate bodies such as universities, hospitals, charitable foundations, research councils, government departments, quasi-nongovernmental agencies, and industrial firms.

The policy of *privatizing* some of the research establishments in the public sector [§4.2] does not seem to be an essential feature of 'steady state' conditions. It may well seem politically desirable to cut taxes and solve the resource allocation problem at a stroke by putting on to the private sector the responsibility for funding the research in question. In general, however, the effect is not to create free-standing enterprises, but simply to transfer each establishment as a going concern to a large firm in the private sector. The real question that remains is not whether the research will then be more efficiently managed, but whether research governed by commercial market forces can take account of all the social needs to which it might be applicable. It is easy to envisage a situation where these needs become so insistent that the government would once again have to allocate resources to this sort of research, and set up public institutions to ensure that it is carried out.

Many basic research institutions have a clearly defined, strategic mission. They are chartered and supported financially to investigate a particular realm of nature, to seek understanding of a particular disease, to explore the basis of a particular generic technology, or to study a particular aspect of the natural or social environment. Institutional targets and capabilities are thus determined principally by external needs and financial resources, and the 'R&D management' model of private industry is often followed.

Where the research is essentially exploratory, a much looser structure may be appropriate. Thus, for example, the scientists employed by the Imperial Cancer Research Fund Laboratories all do research related in some way to cancer. But they are not organized hierarchically into regular research units. They are given the means of working individually or together on research projects that they may have thought of for themselves, provided that these promise some scientific or medical progress in this wide field.

The situation for a multifunctional, multidisciplinary institution such as a university is much more complicated. The general mission of 'advancing knowledge' offers little guidance to institutional policy. The parallel missions of education and public service are only distantly related to research. Funds come from a variety of sources, often with scarcely legible earmarks attached to them. Indeed, what we now call 'academic' science – the systematic pursuit of scientific research in institutions of higher education – is essentially a *traditional* activity, without a compelling rationale. This tradition originated less then two centuries ago in Germany, and although it has been widely copied elsewhere it is still not fully established in a number of advanced countries.

A striking feature of the transition to a 'steady state' regime is that this tradition is being pulled up by the roots to see how it grows. It is not that the holders of the purse strings – that is, politicians and government officials – doubt that research should be done in universities. On the contrary, as the ground landlords of the public higher education system, they often echo the academic claim that good research is the essential adjunct and touchstone of educational excellence [§6.9]. But they insist on looking at the connection much more closely and asking awkward questions. They want to know whether they are getting value for money, whether resources are being allocated in accordance with national priorities, whether the research is performed efficiently, whether highly qualified personnel are being deployed to best advantage, and so on [§5.3]. In other words, the characteristic 'steady state' demands for 'accountability', etc., are imposed directly on the institutions through which funds flow into research.

Now, as we have seen [§5.4], research activity is so finely divided up into disciplines and specialties that even quasi-economic questions such as these are almost impossible to deal with directly. Each question could only be answered after very detailed peer review of the performance of numerous individual research entities. It is no good asking the Vice-Chancellor, President or Rector of a large modern university to take managerial responsibility for the way that, say, a professor of physics is carrying out the study of the superconducting properties of layered copper tellurides, for which a particular grant has been awarded by a particular research council. The objectives of the project will almost certainly not have been a matter of institutional concern in the first place, and the institutional facilities and services that are utilized will mostly have been common to many other projects, possibly in a number of different academic departments [§3.6]. A central university administrator may be able to report to a funding body that the terms of a grant were legally satisfied, and that experts were of the opinion that it was a sound piece of work making a contribution to the subject. But this is rather less information than the funding body would be able to discover for itself in direct communication with the grant holder!

For this reason, any attempt to assess and influence directly the research performance of a whole university is misconceived. An aggregated index of 'quality' is essentially meaningless. The real information, and the real financial leverage, has to be derived from and applied to the many separate departments and research entities located inside the institution.

This is clearly demonstrated by the 'University Research Performance Indicator' exercises of the successive funding councils for British universities. In outline: the block grant to each university contains an element proportional to the sum of the performance indicators determined for each of its 'cost centres'. Thus, the Physiocracy Department at Camford, with its two Nobel Prize winners and its ten Fellows of the Royal Society, publishing hundreds of highly cited papers a year, gets a score of 5, which will gain the university a million pounds; the corresponding department at the University of Ambridge, with one undistinguished professor approaching retirement and trying to encourage his six lecturers to get their work into print, is lucky to get a score of 1, which is too low to get the university the hundred thousand or so pounds desperately needed just to keep its laboratory going.

The Principal of the University of Ambridge may thus get some satisfaction from the fact that her average score is 3.84, but the real question is what to do about the ailing Department of Physiocracy. Should it be closed down altogether, or would a substantial investment in new staff and equipment raise its performance? The persons with the greatest stake in this will be the staff of that particular department. But they already have every incentive to apply for more project grants to support their research. The final effect of these elaborate and time-consuming 'performance indicator' exercises should not, in principle, be very different from what would be achieved by the competition of research entities in their national project markets [§5.11]. In practice, a fully decentralized market system would be less cumbersome and more efficaceous than a centralized command structure.

6.2 The institutional marketplace

Contemporary circumstances are pushing universities more openly into the role of 'holding companies' or 'conglomerates' for the numerous small or medium-sized enterprises that perform research in numerous disconnected fields [§5.10]. This is not really a new phenomenon: absence of an overarching intellectual policy (except the pursuit of truth, etc.) is part of the great tradition of academic freedom. But it is more of an embarrassment nowadays, when an organization is considered unsound and ill-managed unless it has a coherent 'corporate plan' directed towards a few simple goals.

To be honest, however, what universities ought to say is that their only serious corporate goal under the heading of 'Research' is to maximize the income they can gain through backing the diverse research activities of their academic employees. Some of this income is earned directly, through research grants and contracts: some of it comes very indirectly, through the enhancement of teaching quality by association with research [§6.9]. Some of the benefits are short run, such as access to regional, national and international sources of research and consultancy money: some are long run, such as the research funding won through a world-wide reputation for high scholarship. All the rest of the 'plan' – the recruitment of good researchers, the provision of facilities and services, the encouragement of individual and group initiatives, the coordination of research efforts across disciplines, the establishment of new laboratories and centres, etc. – is highly diversified detail, usually of very local and specialized incidence.

To an aristocratic institution such as Oxford, this is tantamount to reminding a duke that he is in the real estate business. But this has long been the familiar reality of even the most distinguished of American universities, whether private like Harvard or public like the University of California at Berkeley. The price they have to pay for corporate independence – and eventually for academic freedom itself – is vulnerability to government policies for the funding of research. This is the significance of current conflicts over accounting for the 'indirect' costs of research [§3.4]. The amounts in dispute or under negotiation – typically, differences of 10–20 per cent on research grants totalling tens or hundreds of millions of dollars for a major research university – are worth a serious fight, which can easily spill over into the courts or the legislature.

Universities and other not-for-profit research institutions are thus developing into corporate participants in an *institutional market*. In effect, they are now expected to make part of their living by publicly selling research services. In the early 1970s, Lord Rothschild proposed that all quasi-academic research establishments in the public sector in the UK should be officially cast in the role of research *contractors*, competing directly for part of their funding from governmental and industrial *customers*. This customer–contractor relationship, for all its deficiencies in practice, has become the design principle for research support throughout the higher educational system.

A conventional market system is thus emerging, where the vendors are universities, quasi-academic, quasi-nongovernmental establishments and private sector 'consultants', and the customers are research councils, government departments, charitable foundations, commercial firms and (increasingly) international organizations such as the European Community. The commodity is essentially a technical service, or, rather, a heterogeneous array of services corresponding to the diversity of specialized research entities within each institution.

Although the actual competition between institutions is distributed across a large number of more or less independent project markets [§5.11], it is a genuine force as a whole. Institutions that are largely sustained (at least in their research activities) by the income they thus gain now appreciate that they are in a fiercely competitive marketplace. A great deal of effort is put into preparing tempting proposals, negotiating the finer details of contracts, and monitoring their performance. Even though much of this effort comes to nothing, an economic analysis would suggest that it is not necessarily all wasted, in that it has the effect of motivating effort and greatly improving the quality of the work that does, in fact, get funded. There is no need to repeat these and other general arguments [§5.9] favouring market competition as a means of allocating resources and getting the best value for money.

In this connection, it is interesting to observe the way that British universities are reacting to the transition from the quasi-command system of funding that had developed under the University Grants Committee since the mid 1970s. Their senior managers are finding that the uncertainties of the marketplace, where at least their successes and failures can be attributed to their *own* past decisions, are preferable to a situation where the major item in the institutional budget for each coming year – the block grant – was decided at a distance by some *other* group, on grounds that seemed often to bear little relation to the health of the particular institution in question.

They have discovered, also, that an experienced and self-confident vendor in a reasonably stable market system has a much clearer rationale for making savings through improved efficiency, and for investing long term in outstanding staff, capital equipment and infrastructural facilities, than if he were still a subsidiary component of a command structure. In other words – and this is all to the good – individual institutions are once more being given the freedom and incentives to exercise the corporate autonomy that is their tradition and ancient pride. If, as many commentators have remarked, this is a move towards the American model, then so what?

Well, there are many obvious things that could be said to temper this rather facile hurrah for the market approach. It would be absurd to suppose, for example, that a market dealing in such intangible commodities as 'research' or 'education' could ever approximate to 'perfection' in the classical economists' sense. Indeed, the institutional market in research services, as now operated in the UK, is clearly very imperfect in just that sense. One of the ways of analysing the transition to a new regime for science is to look at the origins and present effects of these imperfections.

Taking the classical criteria of market efficiency at their face value, we observe, for example, that certain areas of the national institutional marketplace are very far from level. There are many excellent historical and circumstantial reasons why this should be so: for example, the monopsony exercised at the basic end of the market by the relevant government department, or the advantages to be gained from the continued association of advanced research with higher education. But, as we shall see [§7.5, §7.8], the bumpiness of the ground in some of these areas is not just due to such arbitrary 'externalities', but to factors associated with other, more symbolic, 'market systems' operating within the science world.

6.3 Transforming the organizational culture

The transition to 'steady state' science makes new, more stringent, demands on every institution of the R&D system – universities, polytechnics, government and research council establishments and even industrial laboratories. These demands are felt at every level. Individuals, research entities, academic departments and whole universities are expected to be more economically efficient, more scientifically excellent, and more socially beneficial. They are in public competition, across the nation, around the globe, to demonstrate that they can produce more and better with the same or less.

It is easy to exaggerate the contrast with a previous regime, where scientists supposedly led a life of Larry, unsupervised, permanently employed, and provided with all the goodies they asked for to do their own thing. It never really felt quite like that in the 1950s and 1960s. But the memory of a benevolent post-War era of almost unrestrained expansion [§4.1] still lingers in the academic world, and there is a wistful hope that these fortunate conditions might return. Consequently, there is great resistance to changes in the personal and institutional arrangements that seemed at that time to work so well – especially for those who rose successfully to fame and influence within them. This social inertia is particularly heavy within academic institutions, not only because of the strength of 'tradition' as a principle but also because of the way that power is decentralized and devolved to collegial subgroups and individuals.

Indeed, this loose-fitting, open-ended organizational framework made it easy at first for universities and other academic institutions to absorb the stresses of the new financial and policy environment without radical structural changes. OK, so we're not going to get that new building they promised us: let's set up a Departmental Space-Utilization Committee and see if we can at least get Dr B— to give up one of his three photographic darkrooms. OK, so we are not getting enough students now to justify that Department of Very Unusual Studies that we set up in 1972: we can't do much about the professor, but we'll redeploy the lecturers into other departments, where they will be able to help with the teaching and maybe go on doing their research as before. OK, so there is no tenure-track position for that brilliant post-doctoral fellow: we'll give her another three years as a research assistant on a project grant, until a vacancy occurs. In spite of functional distortions, excessive workloads, and severe administrative strains at all levels, the customary procedures were still followed for a long time.

But the external stresses continued to grow, and the system began to crack internally at a number of points. What we are now seeing, most strikingly in the UK but quite visibly in most other advanced countries, is a genuine process of *restructuring* right at the heart of every institution. This term is often used to refer to a phase of financial retrenchment and managerial redeployment of resources to meet the challenge of a difficult situation. Here we would emphasize the literal meaning of a major change in the working relationships between the subgroups within a social institution and their relative spheres of responsibility and action.

Much of what is said about science policy concentrates on the arrangements by which public funds are allocated downwards to various research areas and research organizations. We have already remarked [§6.1] on the way that these arrangements force universities and other institutions to become much more entrepreneurial in spirit. In their search for alternative sources of funds, some very monastic institutions have had to open themselves up to the world. Industrial research services and technical training programmes have to be tailored to the requirements of their customers, and actively marketed. Intellectual property rights such as patents [§2.7] cannot be expected to earn an income unless they cater to ascertainable human needs. Even if they are not being pressed directly to make their work commercially profitable, researchers and research entities are plunged into an atmosphere permeated by such worldly considerations.

These demands, in turn, affect institutional security and morale. People have to be much more sensitive and responsive to calls for greater responsibility towards society and its needs. This can be very unsettling for research scientists and scientific institutions whose inner moral fibre was the belief that the search for knowledge was valuable 'for its own sake'. It has always been easy for scientists to pay lip service to vulgar utilitarian values, provided that these did not actively challenge the supremacy of the academic ideology. They could always refer the critic to Faraday's discoveries in electromagnetism, for example, and point to the subsequent social and economic benefits of the electrical industry [§2.6]. But such arguments do not cut much ice in negotiating a research contract with a present-day electronics firm seeking to speed up signal transfer between the microchips in a computer.

This significant drift away from notions of 'purity' in the rationale of academic research is directly driven by the general resource scarcity of 'steady state' science. What may be equally significant for science and for society is an even more radical transformation of the *organizational culture* of academic science, as between individual researchers, research entities, academic departments and faculties, and university central administrations. These changes will be our concern for the remainder of this chapter.

6.4 Scarcity, urgency and anxiety

The most obvious of the new demands on research institutions is that they should be *financially* efficient. It may seem strange that this is seen as a novelty. Surely, universities and research establishments have always had to balance their books, seek new sources of income, watch their costs and invest prudently in new facilities. No doubt such considerations become more urgent at a time of general economic retrenchment, when the opportunities for income growth are severely limited and every item of expenditure has to be strictly accounted for and justified. Many academic institutions are very ancient, and can look back over the centuries to moments of crisis, when the total savings they could achieve by normal means fell below the gap between costs and incomes, and all other policy principles had to be thrown overboard to save the ship. But anxiety about financial solvency now seems endemic, and permeates the whole research culture.

Several forces combine to generate this anxiety. Almost by definition, 'steady state' science is a regime of resource deprivation. Research funding bodies inevitably transfer to their clients and customers the budgetary constraints and cuts applied to them by governments. A research council panel [§5.4] that can only fund half of its portfolio of 'alpha- rated' proposals naturally tries to trim each project budget to its absolute minimum, pushing applicants to the margin of feasibility in terms of researcher time, technical staff and experimental equipment. Since many of the services and facilities actually used in research are not paid for specifically by funding agencies [§3.5], systematic internal accountancy of, say, computer time is called for. Again, funding bodies may try to economize by applying strict rules on the types of costs they will allow – for example, excluding any provision for secretarial work, or for travel to attend conferences related to the research. In the course of time, these attitudes become widespread, and are internalized by institutions, so that some expenditures – for example, for academic staff travel – are generally regarded by research administrators at all levels as 'wasteful' or even 'corrupt'.

Another factor might best be described as *ideological*. In the 'enterprise culture', cost accountancy is the dominant discipline, and profit-and-loss efficiency the supreme virtue. As we have seen [§6.2], research institutions are deliberately pushed out of the warmth of public financing and administration into the icy waters of market competition. This has many consequences, not all of which are desirable, but it does have the intended effect of concentrating their minds wonderfully on operating costs. Universities appoint business consultants to take a critical look at their old Spanish customs, and do debate seriously whether, for example, it is necessary for every academic department to have its very own library, or whether it might not be cheaper if all laboratory, workshop and office equipment were purchased and maintained centrally. Indeed, governments wedded publicly to the 'invisible hand' of the marketplace still get so anxious about this sort of thing that they adopt standard *dirigiste* practices, such as 'efficiency audits', to implant these entrepreneurial attitudes more deeply.

There is another more general factor, however, springing from the changing nature of research itself. In the past, science and scholarship were essentially *labour intensive*. The cost of doing research could be attributed in very large part to the time spent on it by tenured academics, research fellows, doctoral candidates, and technical support staff. The bulk of these costs were contractually determined ahead for the several decades of a permanent employee's career, and could not be varied greatly from year to year to meet new demands. The same applied to the maintenance and amortization costs of buildings and equipment. For this reason, both the income and expenditure sides of the balance sheet were customarily kept very stable – for example, by a five-year cycle of block grants to cover a fixed 'staff establishment', where it was permissible to employ a new person when, and only when, there was a vacant post.

As we have seen, however, research is becoming more and more technologically sophisticated, and more and more dependent on elaborate, and expensive apparatus [§3.1]. This apparatus has a very high rate of obsolescence, and has to be written off over just a few years [§3.3]. Labour costs, also, are becoming less stable, with larger numbers of research and technical staff employed on short contracts to carry out specialized tasks [§7.2]. Some of this work may not even be done by institutional employees. In an advanced industrial country such as the UK, it is quite feasible, and often economically efficient, to buy in specialized services, such as computing time or instrument maintenance, or to contract out some of the work, such as the design and construction of one-off instruments, to external firms. Institutional managers and research leaders must consider a much wider range of expenditure options, involving much larger variations of cash flow, if they are to seize the opportunities opened up by a more rapid advance of knowledge on many different fronts.

The financial anxiety that now seems a permanent feature of research activity is not, therefore, generated entirely by greatly increased competition for research funds as such. This may seem the immediate cause of such symptoms of financial aridity as the shortening of grant awards and the increasing employment of temporary research staff. But this anxiety is also due to the speeding up of the research cycle as researchers in every field desperately try to keep at the fast-moving leading edge of their specialty.

There is no decisive evidence that the gestation time for a significant scientific discovery – if that could be given a clear meaning – has actually diminished since, say, the beginning of the twentieth century. Indeed, in 'Big Science' fields such as high-energy physics this period of gestation has probably increased, because new research facilities take so much longer to build than they used to. There is also a feeling that the managerial practice of chopping research into self-contained projects with stated objectives [§5.8] obscures the long reach of the many research programmes that are still conceived, and carried out over decades, by dedicated researchers. But the hustle and bustle of electronic mail, proliferating research literature, publication and citation indicators, and frequent international conferences, creates a general atmosphere of urgency, to which financial instabilities and anxieties seem a normal accompaniment.

6.5 Financial efficiency

Slowly but surely, external financial pressures can be relied on to produce beneficial administrative rationalizations, such as streamlining internal committee procedures, sharing apparatus between departments, and introducing more business-like infrastructural systems for secretarial services, stores, technical facilities, etc. R&D institutions tend to become very set in their ways, and often provide protected niches for bizarrely incompetent organizational actors. The time and effort that goes into such exercises should eventually pay off in a measurably improved supporting framework for research [§3.4]. There may also be an unexpected dividend from challenging and changing customary practices and personal roles in such an administrative reform.

On the other hand, budgetary pressures push institutional managers into shortening their accountancy cycles – for example, by employing temporary research staff on shorter contracts or even by not settling outstanding bills with suppliers until the last moment. Such manoeuvres may benefit the cash-flow element in the institutional accounts, but they may also reduce the overall efficiency of the system by frustrating personnel policies and the long-term planning of internal resource allocations. As we have noted, the new atmosphere of hustle and bustle veils the continued reality of scientific research as an activity reaching to distant horizons, with a characteristic time-scale measured in decades rather than months.

For all its stateliness, the UK system of quinquennial block grants [§6.4] was well matched to this time-scale, although it was not realistic to suppose that universities would spontaneously move themselves along (even at that rate) without other, more compelling, external stimuli. This system actually broke down because institutional budgets are dominated by the salaries of tenured research staff, whose pay awards in a period of high inflation caused very severe problems. Nevertheless the transition to a 'steady state' regime would have been less distressing and damaging if it could have been set goals and firmly steered within a less hectic financial atmosphere.

One of the general objectives of science policy is to achieve major economies through specialization and the division of labour between research institutions, nationally and internationally [§3.9]. But market selectivity cannot operate effectively under conditions of extreme financial stringency. Instead of spurring scientists on towards exciting new goals, the effect is to produce organizational rigidities which prevent the adoption of long-term policies for institutional and personal development, for the maintenance and modernization of capital equipment, and for programmes for scientific advance over a broad front. Worse still, institutions find it difficult to build up reserves to exploit serendipitous discoveries, or to support very novel work on a small scale. They thus become more enmeshed than ever in the bureaucratic trivialities of research funding agencies, which also ought to be thinking and acting on a broader scale.

Another consequence of undue concern about financial efficiency is to put disproportionate pressure on the funds allocated by institutions to the 'marginal' facilities for research, such as libraries, travel funds, secretarial assistance, etc. This tendency is enhanced by top-down accountancy procedures which emphasize performance indicators of efficiency in terms of quantitative inputs and outputs, with little allowance for incidental but necessary expenditure on administration, technical support services, buildings, instrumentation, infrastructural facilities for communicating, storing and retrieving research information, and so on. The growing importance of items such as these is characteristic of the increasing collectivization of science [§3.4]. These *indirect costs* thus add up to an increasing proportion of the research expenditures of universities and other research institutions.

Unfortunately, as any professional accountant will tell you, such costs have to be paid for in hard cash; yet their benefits are usually so widely shared and spread [§3.6] that they are only notionally attributable to those who gain from them. They are also, at the margins, very intangible, shading off into public relations and semi-official perks. It is scarcely surprising that institutions and funding agencies are locked deeply in conflict over how such costs should be defined, determined, aggregated over many different activities, sub-divided proportionately to more direct costs, and added into project grants and research contracts [§6.2]. In the United States, where this conflict has been brewing for more than twenty years, it has now blown up into a major political storm, with accusations of malpractice and injustice between federal legislators and university authorities. In the UK, the government is trying desperately to use its prerogatives to keep the whole issue under control, even though this is now beginning to occupy more and more of the negotiating resources of civil servants and senior academics. In other countries, where academic science is funded in rather different ways, this issue may not have emerged yet into the light of day, but one can be quite sure that it is causing acute concern in science policy circles.

The fact is that this is not just a two-sided conflict between the universities on the one side, and funding agencies on the other. Individual academics and leaders of research entities have very mixed feelings on the subject. In practice, they often have a much more direct and active interest in maximizing their operations in the project market of their specialty [§5.11] than in the prosperity of their university. Indeed, they are strongly motivated to sell their services as cheaply as possible, resenting the 'additional' indirect costs that need to be charged to the customer if the institution is to remain solvent. They do not care whether, say, a grant from the European Commission returns much less to their institution than an industrial contract to do the same research, and they moralize eloquently on the desirability of providing research services to a medical charity at what is now far below their real cost. Apparently arbitrary administrative variations in the incidence of these charges not only distort competition in the institutional and project markets: they are significant historical markers of the transition from 'patronage' to 'contract' in the political economy of science [§4.2].

Longer experience of strict financial constraints may teach institutions how to plan their expenditures so as to avoid some of these harmful effects. They may even learn some less admirable devices, such as 'moonlighting', 'creative accounting', jobbing backward in official plans, and selling the same project to several customers. This in turn may induce harsher procedures of external audit. The unwritten conventions and traditional courtesies of academic life may be deemed a luxury in this more combative culture.

Continual awareness of resource limitations almost defines 'steady state' science [§4.7], and will surely remain a permanent feature of the research system. Apparent inefficiency in matching means to ends will not be tolerated. Nevertheless, conventional accountancy criteria are not really applicable to such a speculative activity as curiosity-driven research. Science is an industry where obvious success is rare and elusive, where the duplication of an investigation is not necessarily wasteful, and where the eventual profits seldom come to the original investors.

6.6 Is all science simply being bureaucratized?

'Steady state' conditions accentuate the need for institutional *management*. A university or research laboratory under severe financial pressures, in a rapidly changing scientific and technological environment, with a searching evaluation of its performance in prospect, cannot be run as if it were a cooperative of individuals or small groups cheerfully following their own bent. As we have seen, intense pressures at the top of the research system force hard choices: these pressures are transmitted down through institutions to the level of each research group.

Very difficult, but very firm, decisions may have to be taken. It may be necessary to close down a viable research unit, or even a whole establishment. An agreement may have to be negotiated across several faculties for a major investment in facilities and buildings. Dedicated researchers may have to be taken out of their accustomed fields, assembled in a multispecialty team to exploit a novel opportunity [§3.7] – and then redeployed if the new project does not seem to be paying off.

It is all very well for a university to formulate an impressive corporate plan, where it promises, say, to diversify its research funding, collaborate more closely with local industry, or be more aggressive in marketing its scientific services. In practice, this will require administrative action to divert the energies of many members of academic staff into unfamiliar channels. The whole organizational culture of science has had to be dynamized and tightened up.

Top-down decision-making is not, of course, a novelty in the scientific world. From the end of the nineteenth century, a number of industrial and governmental research organizations were set up along conventional bureaucratic lines, with a formal hierarchy of directors, managers, laboratory heads and group leaders [§4.2, §7.4]. Since the Second World War, these organizations have expanded enormously in size and numbers. In an economically advanced country they now employ the majority of the full-time research scientists and technologists. Generally speaking, they have also become even more 'managerial' in style, although their informal practices are still sometimes reminiscent of more individualistic 'academic' models [§5.9]. The self-conscious art of 'R&D management' simply takes for granted a top-down command structure, designed to fit into the corporation or agency they serve and to carry out its will. The immense R&D operations of industrial firms such as IBM, Shell, or Ciba-Geigy, of government departments such as the US Department of Energy or the UK Ministry of Defence, of international organizations such as the European Community or CERN, of national research councils such as the French CNRS (Centre National de Recherche Scientifique) or the Australian CSIRO (Commonwealth Scientific and Industrial Research Organization), could not really be administered and directed in any other way.

It was on the cards – and many people assume that this is what has happened – that all scientific research would be restructured along these lines. Strong internal government by a bureaucratic hierarchy seems the rational response of an institution faced with the challenges of 'steady state' conditions. Certainly, many practices from the world of R&D management have been imported into academic science. Traditional collegial procedures have been streamlined, and transformed into more formal managerial structures, where consensus is no longer the condition for change. The decisions that need to be taken are often too painful for negotiation through committees of those who would be affected by them. The responsibility for difficult choices has to be pushed upwards to more senior authorities.

Until perhaps twenty years ago, the term 'management' scarcely appeared in the university or civil service vocabulary. Now it is considered such an essential component of scientific activity that systematic management training is required to fit scientists for senior posts. In our ancient universities, where even 'administration' was previously considered a demeaning occupation for a scholar and a gentleman, the professors are now designated 'middle managers', whose departments are 'cost centres', where they are responsible for the 'work programmes' of their 'personnel'. Some of this is just fashionable jargon, but it bears witness to significant changes right down into the roots of the research culture.

What's so strange about that?. Isn't that the way that everything is organized nowadays: industry, government, religion, war? Why not simply accept that science and education have been 'industrialized' or 'bureaucratized', along with all the other collective activities of our society? This view has often been expressed (usually regretfully) by thoughtful social historians of science [§1.3], and has much to be said for it in general terms. For example, it makes explicit the increasing incorporation of the scientific enterprise into the power system of society. To the extent that academia itself is being integrated into a powerful military–industrial complex, so it must follow that academic research is being reorganized along military–industrial lines.

It would certainly make the present book easier to write – perhaps even superfluous – if the radical changes occurring in the research system precisely followed from and copied changes in the surrounding society. In reality, the situation is more complicated – and more interesting. As we have already observed [§5.9], the arrangements made for funding academic science nationally do not imitate military or industrial command structures. There are few counterparts, in other publicly funded bodies, of the organizational autonomy permitted to universities, or of the powers actually exercised by specialist peer review panels in funding agencies.

The transition to 'steady state' conditions has not been marked by wider integration under firmer managerial controls: at the academic end of the spectrum, the movement has been in the other direction, towards sharper organizational segmentation, and fiercer competition for resources, in more 'level', better framed, markets. This rejection of outright *dirigisme* applies as much to the detailed organization of scientific work within research institutions as it does to the research system at large.

6.7 Franchises for research entities

It would take us too far afield to describe and analyse all the structural changes that are now taking place in the internal governance of academic institutions. Only some of these are directly associated with the bounding of resources for research or with the increasing 'collectivization' of science. In countries such as Britain, a much more important factor has been the massive expansion of an elite system of higher education at a time when the total educational budget was being capped for economic reasons. In every country, universities seem to have the same educational responsibilities, yet they are very varied in the way that they are governed, and are responding in very different ways to these pressures.

As far as research was concerned, the traditional academic decision process should not be dismissed as ineffectual. It may have been lengthy, but it was often far from cosy or permissive. Tough internal bargaining over the allocation of resources has always been a familiar feature of universities and other scientific institutions [§7.3]. Indeed, the notorious academic phenomenon of interminable political in-fighting over staff appointments and promotions shows the importance attached to individuals as long-lived sources of research capacity for the institution.

To put this the other way round: the tradition was that universities took very little corporate interest in research, apart from a great deal of initial effort to select staff of the highest possible research reputation or promise. Once they had been given tenured appointments, it was up to these staff, as individuals, to make their own way in the scientific and scholarly world. The basic principle in academic science, in the research councils as well as in universities, was to treat a scientist who had reached this position as a 'prime mover', whose research activity, whether alone or at the head of a group or unit, should be encouraged and supported on its own merits, regardless of how this related to the work of others in the same institution [§7.7]. This ethos of individualism shows up clearly in the established policy of some research organizations, such as the UK Medical Research Council, of closing down a research unit when its director retires, to open the way for new enterprises, by new leaders, with new scientific ideas.

In most universities, there remain many vestiges of a primitive structure of independent scholars, doing 'their own work' as researchers [§7.8], and loosely grouped into self-governing 'faculties', 'departments' or 'chairs' for teaching purposes. Nevertheless, research in the natural sciences has long required collective facilities and services [§3.4]. Rather more definite administrative arrangements had to be made to generate, develop and manage these facilities on behalf of the groups of academic staff who were to share in their use. In British and American universities, these arrangements crystallized around academic 'departments', defined and differentiated in terms of their 'disciplines'.

For the best part of the twentieth century, the principal structural unit in academic science has been The Laboratory, or The Institute – not just a specialized place for doing research, but a many-sided social organization. Wander around the campus of any major university, and you will see large buildings named after wealthy benefactors, eminent scholars or powerful academic worthies. A closer look reveals that each of these is the site, the home, the citadel, of a scientific discipline. The abstract subdivisions of the map of knowledge [§3.8] take material and social form in the physical layout of the campus.

As a prime example, consider the Cavendish Laboratory at Cambridge. Its name celebrates Henry Cavendish, the reclusive scientific genius, along with the aristocratic family that endowed the original building more than a century ago. The modern building with this name stands a mile away, and houses the whole Physics Department of the University. In addition to rooms littered with the paraphernalia of experimental research, it has offices, lecture theatres, seminar rooms, store rooms, workshops, computer terminals, a library, common rooms, a cafeteria. Hundreds of people – professors, lecturers, research students, technicians, secretaries, administrators – spend all their working days there. Hundreds of undergraduate students come and go hourly, to attend lectures and laboratory classes, since this is also the geographical location and organizational centre for physics teaching in the University of Cambridge.

Talk to the people who work in such a place, and you will find that it is almost a world of its own. Apart from receiving their pay cheques, they may have little personal connection with 'the university'. Their working conditions may be laid down formally by some university committee, but are largely determined in practice by negotiation with the senior staff of the laboratory. Although the financial books of the department may be kept in the central accounts office of the university, they merely record transactions decided almost entirely at departmental level. In the golden years, from the late 1950s to the early 1970s, [§4.1], there was always enough money to purchase minor equipment and consumables. If anyone retired or died, the vacant post could be filled as a matter of right. The main task of a head of the department in relation to the university seemed to be to wangle as large a share as possible of the new money coming in for academic posts, buildings, support staff facilities and research equipment. Having won the funds, the department then had almost complete discretion on how they were spent.

The style of governance *within* an academic department can vary greatly, depending on factors such as size, subject, operational complexity, general national or institutional practice, the status of more senior staff, or the personality of its head. The systematic managerial arrangements required for a large research laboratory in an experimental science would be out of all proportion for, say, a modern language department of a dozen professors and lecturers working mainly on their own. There is a notorious tradition of tyrannical professors lording it over a sorry bunch of underlings. There is an equally solid, though less publicized, tradition of discreet leadership by a *primus inter pares*. Some great scientific laboratories have been run entirely collegially, as if they were kibbutzim: others, just as successful, have been ruled as autocratically as commandos. In some countries, all the academic staff working in a broadly defined discipline such as chemistry are deemed to be members of the same department: in others, such a discipline may be differentiated formally into a number of independent 'Chairs', or 'Institutes', whose staff uneasily share the same building and research facilities.

Nevertheless, this diversity occurs within a common pattern. Over the past 50 years or so, academic science in most countries has crystallized around university teaching departments. Whatever their internal strengths and weaknesses, these have become the nerve centres of research activity, for the individuals that belong to them, and for the institutions to which they belong. The question is: how has the transition to 'steady state' science affected this basic pattern?

The general effect has been to reduce the research responsibilities of disciplinary departments, upwards and downwards within the institution. On the one hand, it seems necessary to tighten up the chains of command, and bring the departmental baronies more closely under the control of the central university administration. This means more than harmonizing and rationalizing services and facilities, getting departments to cooperate in new research programmes, or winning acquiescence to radical surgery of faculty structures. It means penetrating departmental defences to give the central authorities more direct influence over the research activities of individual scientists and research groups – for example, in the wording of research contracts, or the ownership of intellectual property rights [§2.7, §7.6].

On the other hand, large academic departments are usually sub-divided, formally or informally, into 'research entities' that can no longer rely on departmental funds for much of their work, and have to go out and seek their fortunes in highly competitive specialized project markets [§5.10]. Not surprisingly, they are most reluctant to subordinate themselves administratively, or surrender any part of what they regard as their hard-earned income, to the department or the university. It is natural for institutional subgroups – research entities within a university department, departments within a school or faculty – to resist the loss of autonomy to higher organs of the institution.

It should be said, however, that a research unit or department designated as a 'cost centre' is made more aware of its identity and potential independence of action, even though it 'belongs', legally and financially to the parent institution. Its response may be to tighten up its research programme, and market its products more effectively in the more competitive internal environment of the institution, as well as in the outside world.

The internal fragmentation of departments is accompanied by the growth of a network of interdepartmental connections. The study of a complex problem requires expertise and technical skills from a number of different points of view [§3.8]. It is becoming more usual for a research project to be proposed and undertaken as a joint enterprise of researchers from several different departments – perhaps from several different universities. National funding bodies now encourage and support the establishment of quasi-independent multidisciplinary 'research centres', each with a research programme focussed on a particular technological or social problem area, such as engineering materials, AIDS, science policy or urban renewal [§5.8].

Whether organized departmentally by disciplines, or as multidisciplinary research centres, academic research units are also being encouraged to establish direct links, both cooperative and competitive, with other similar units in other institutions. It is not unusual for a project to be proposed jointly by researchers in cognate departments in several distinct universities, sometimes even in different countries [§3.6]. These working linkages are not confined to the university or research council sector. As they extend into the private sector, they create a multitude of bonds between individual academic research groups and industrial firms. This detailed interpenetration through a variety of research contracts, commercial agreements, exchanges of personnel, etc. is probably the most effective way of facilitating the technological and commercial exploitation of basic scientific discoveries [§9.7].

Such developments are highly desirable from many points of view, but they generate serious tensions in the strongly marked departmental structure that had become established in academic science between and after the World Wars. By extending 'horizontally' across the whole research system rather than 'vertically' within each institution, these networks of collaboration and mutual obligation hinder inter-institutional competition in project markets. The demand for more centralized management of universities runs counter to the need for funding agencies to take more responsibility for national research capacity in specialized fields. It is not at all clear how such 'matrix' issues, which are inevitable effects of the increasing 'collectivization' of science, will eventually be resolved.

The fact is that universities and other academic institutions still find great difficulty in acting as independent corporate bodies, each with an integrated structure and a coherent policy. They are highly segmented administratively into teaching departments and research entities. Institutional control over these subunits is constrained by national coalitions organized around educational disciplines and research specialties. The transition to 'steady state' science has not favoured increased centralization and managerial integration. Indeed, in economic terms, large institutions are often treated as bankers, or distant holding companies, by trans-institutional coalitions of their staff members competing as relatively small firms in the project markets of their respective disciplines [§5.10].

One response to this feature of academic life is deliberately to encourage the managers of research entities to act as independent entrepreneurs *within their own institution*. Administrative mechanisms are devised that force them to compete directly with one another in a so-called *internal market* for institutional resources, such as technical services and funds for new research equipment. In effect, it is a *franchise* arrangement, where the research entities are independent entrepreneurs in the external marketplace, but have to pay rents to their parent institution in return for technical services, senior personnel and access to venture capital. But this is only one of the many novel ways in which large corporate bodies nowadays manage the financial and administrative links with their subsidiaries.

As time goes on, universities and other research organizations in the science base will presumably evolve managerial practices and procedures that are consistent with both their historical missions and their future functions. They may also learn something from the ways that scientific discovery and technological invention are actually managed now in many industrial R&D organizations – ways that are often reminiscent of the academic tradition of small, more or less independent, teams, led personally by talented researchers, in a generally supportive administrative environment where people are trusted to work together on agreed problems for the common good.

6.8 Selectivity and specialization

'Selectivity' is one of the buzzwords of the new regime in science. It is applicable to a wide range of situations, from choosing regretfully between two alpha-rated research projects to closing down – even more regretfully – a whole scientific establishment. To those who exercise it, it suggests thoughtful consideration of all the options, leading unavoidably to hard choices between almost equally worthy claimants. To those on whom it is exercised, it suggests the triumph of managerial elitism over anarchic egalitarianism.

What selectivity has come to mean in practice is a systematic policy of *concentrating* research activity into a smaller number of more *specialized* units. The purpose is to gain the benefits of the 'division of labour' and of 'economies of scale'. Some of these benefits [§3.9] are not as real, or as easily obtainable, as is generally supposed. Nevertheless, this notion is widely accepted, and animates many current policies and practices, with radical institutional consequences.

The contemporary advocates of 'selectivity for specialization' do not always seem to appreciate that it is nothing new in science. Institutions have always been aware that their research achievements will be a prime factor in the future inflow of students, staff and research resources. They have always sought to gain these advantages by very careful, highly competitive, selection of their academic staff. Appointments and promotions have always been fought over and won on the basis of achievement or promise in the international marketplace of science [§7.3]. Individuals, competing for such appointments, have almost always followed the path of extreme specialization as the securest way of demonstrating this quality. The Matthew Effect [§3.9] then comes into action, and accounts for all the students, research assistants, academic colleagues, institutional resources, external grants, patronage, etc. attracted into the sphere of operations of an eminent scientist. Taken together, these factors are sufficient to ensure the concentration of research activity into specialized centres.

How is it, then, that the intensification of this natural process has recently become a major policy goal? To a very large extent, this is merely a symptom of the transition to a more collectivized, more competitive, yet financially bounded scientific enterprise. Scientific, technological, industrial and commercial factors combine to force the pace of change. In many fields, there has been a substantial increase in the critical mass of effort required for competitive research [§3.6]. Centres of scientific excellence in new specialties have to be built up to this size more rapidly than they would if left to grow through the normal workings of the Matthew Effect. Total resources of money and people are limited, so they must be transferred systematically from activities that are in decline. 'This decision hurts us as much as it hurts you, but do we have any alternative . . . (etc., etc.)?'

Selectivity actually operates at many levels. It is implicit, for example, in the publication of a list of long-term national priorities for science [§5.7]. The writing is on the wall for research entities working in low-priority fields, even when this work is of high scientific quality. The cry may then go up that the nation is losing the capacity to respond rapidly to possibly important developments in those fields [§8.9]. 'Do you realize, Minister, that if our Atomic Super Induction Laboratory is closed down we shall be in no position to compete with the Japanese if they invent a Super-Inducting Atomiser . . . (etc., etc.)?' Such policies obviously have their risks, as well as their benefits. The prospects *of international* selectivity for *national* specialization in science and technology will be explored in chapter 8.

Choices have to be made between allocating resources to infra-structural facilities and funding specific projects. In 'Big Science' [§3.1], these facilities generate large and indivisible budget items, which are not easy to balance against long lists of small items of 'little science'. This problem, again, is confounded when international programmes are involved [§8.6]. 'But Chairman, our share of the cost of running the Multinational Multifunctional Super-Inductor for a week would pay for twenty post-doctoral fellows in Differentiated Microbotanology for three years . . . (etc., etc.)'

Even within relatively 'little' science, a balance has to be struck between projects proposed by teams of researchers and proposals for more individual research work. Although the trend in 'steady state' science is strongly towards collaboration and team work, there are still many fields where effective research can be done by competent scientists working alone on projects of their own devising. Should review panels favour such lone hands – especially when they are located in departments that are otherwise almost barren scientifically – or should they systematically foster collaborative projects that might nucleate new interdisciplinary research centres? The administrative cost of making large numbers of small grants is quite considerable, and yet the first priority in some cases is for very modest resources for communication, secretarial and other networking facilities. 'Look, he is only asking for £5k, but could he really do anything worth while for such a small sum . . . .(etc., etc.)?'

These are just some of the ways that 'selectivity' emerges as a policy consideration in the deliberations of government advisory bodies, research councils, and specialist peer review panels. The bodies that allocate resources to research cannot shrug off their responsibility for the *future* state of the research system, and its capacity to meet future demands. Systematic action has to be taken not to waste resources on research entities – large and small – that do not seem likely to have a place in that future. Means have to be devised to redistribute these resources – per-sonnel as well as funds and equipment – into new entities, new facilities, new specialized centres, that are strong enough to be competitive in world science.

Outside academia, this process is difficult but manageable. Despite the quasi-academic atmosphere that they have assiduously cultivated, government and research council establishments have always been subject to major exercises of reorganization and restructuring. What is new is the 'steady state' constraint, which means that the necessary resources and personnel may have to come from or go to other units in the same field.

At the start it is easy enough to identify notoriously poor performers. Yet it may still require considerable managerial fortitude to cut hard into a large institution enclosed in the hard shell of a great history. In recent years, institutions as eminent as the Royal Observatories have had to suffer this indignity, although not without a great deal of private lobbying and public outcries. In due course, however, the rotten wood runs out, and living branches have to be pruned. Apparently arbitrary choices have to be made between efficiently run institutions of high repute. It becomes necessary to discriminate between actively advancing research fields of high promise.

This process, however diplomatically it is undertaken, is bound to generate organizational turmoil. The initial process of assessment puts into question the achievements and prospects of all the institutions and people involved. There is no way of evaluating scientific performance [§5.5] that does not leave room for strenuous argument as to whether justice has been done [§9.4]. Even where there is good reason and quite sufficient managerial authority to make clear choices, these will be resisted vehemently by those people not favoured by them. Such situations are almost impossible to handle without violating the conventions of personal consultation and consensual action established over long periods of scientific growth.

Within academia, with its ancient traditions of individual autonomy and institutional self-government, it is almost impossible to be drastically 'selective' without support from outside sources of power. But universities everywhere are very resistant to external pressures. In the UK, for example, they are legally independent institutions, nominally of equal status, receiving block grants of public money for education and research. In the past, they regularly exercised their freedom to undertake teaching, advanced study and research in any field of science, regardless of similar decisions by other universities. Although national funding bodies have effectively curbed this freedom in recent years, they have no authority to instruct institutions what they should do about such matters in detail, and have no formal means of enforcing such instructions if they are not carried out.

Indeed, as we have already noted [§6.2], the tendency at national level has been in the other direction, away from 'command' structures towards 'market' systems for the allocation of resources for research. But bodies such as the UK research councils, who now pay most of the direct costs of academic science, are becoming more aware of the institutional impact of their award or non-award of grants and contracts. They are beginning to learn that, as long-term customers for research, they should not be funding projects entirely on their scientific and social merits [§5.4]. A peer review panel specializing in, say, subconducting monomerization has no right to complain that there are no alpha-rated projects being proposed in this new field, when, as it happens, their own adverse decisions of five years before had caused the only good British researcher interested in developing this subject to pack up and go to Australia. It may well have been that his earlier proposals seemed of distinctly lower scientific quality than the other work they were funding at that time; but they should have seized the opportunity of supporting the build-up of a British centre of expertise – and eventually of excellence – in this new field.

The trouble is that classical economic forces only work in the long run, by which time, as John Maynard Keynes quipped, we are all dead. In principle, customer selection for quality in perfectly competitive markets for research proposals should achieve precisely the optimal degree of concentration and specialization: in practice, these markets are very imperfect, and work much too slowly to be effective in the rapidly changing scientific, technological and economic environments of our times.

This, presumably, is the thought behind the elaborate procedure that has grown up recently in the UK, whereby a substantial proportion of the funding of higher education depends directly on formal assessments of research performance [§6.1]. But this has turned out to be a very messy and unsatisfactory business which has not yet settled into an established routine. Not surprisingly, a procedure whose outcome can spell life or death for research activity in a thousand different university departments generates intense anxiety and controversy. Beyond that it has attracted serious technical criticism in almost every aspect: the assessment criteria, the data going into the quantitative indicators, the composition of the peer review groups, the classification of 'cost centres', the role of non-researching academic staff, weight factors in aggregating sub-indicators, the effects of size, and so on, almost *ad infinitum*.

Scientific policymakers in other countries probably have much the same goals, but (not surprisingly) are not hurrying to follow the UK along this path until they see where it finally leads. Indeed, in a free enterprise economy such as the United States, such a policy would be anathema. What we can see already is that the political and economic pressures for greater 'selectivity' in the running of a national research system, genuine as they are, are not easily translated into an effective policy. The underlying *dirigiste* rationale is questionable. The notion of 'excellence', for example, has so many different dimensions that it cannot be summed up in a single numerical indicator, valid for all managerial circumstances and seasons. So much depends on the peculiar capabilities of individuals and small groups, continually adapting to changing intellectual and institutional environments.

In the end, selection through the 'invisible hand' of competition in project markets looks more equitable than any attempt to steer institutional change from the outside, and may prove just as effective. Policy for 'selectivity' might do better to concentrate on opening all aspects of the research process to this type of competition – for example, by applying it directly to the scientific work of all tenured staff in universities and other research institutions. The academic tradition takes pride in the protection that it gives to the promising young scholar with a life-long project to bring to fruition: it has often been used as a comfortable shelter for middle-aged people who have run out of scientific steam but who continue to use scarce resources to little purpose. The abolition of unquestioned permanency of employment in UK universities is a step in this direction [§7.2]. It will be interesting to see whether 'steady state' pressures eventually produce a similar change in full-time academic research organizations such as the French Centre National de Recherche Scientifique, where the principles of tenure and individual scientific autonomy are still deeply entrenched.

Nevertheless, there is no escaping the overall logic of the situation. One way or another, a process is in train that must inevitably transform the relative roles and functions of universities and other institutions, and the relative roles and functions of academic departments, research entities and individuals. The systematic channelling of research funds into a selected research unit, with the avowed purpose of building it into a centre of excellence, is not a new process. In the past, academic units have also had to close down, or be merged with others. But these changes took place so slowly and infrequently that they did not threaten the established order: the institutional stratification of esteem, equality within a professorial oligarchy governing a hierarchy of academic staff, egalitarian distribution of research resources to departments, collegial decision-making, and so on.

'Selectivity' works best when it extends and intensifies this competitive tradition of science. Public reputation was always, and is still, a very potent force. Open competition between institutions cannot, of itself, generate scientific creativity and innovation, but it encourages the employment of creative and innovative scientists. An environment of vigorous, orderly and fair public competition between research entities is the necessary condition for achieving all those objectives that are bundled together under this heading. Selectivity implies quality assessment, the elimination of unproductive research, specialization, and the natural concentration of capacity into self-organized centres of excellence. But this requires that resources and responsibilities are devolved in a way that fosters *self-selective* excellence, and gives institutions the autonomy and flexibility they need to *demonstrate* this excellence and to undertake radical change.

Moreover, if the selection process is serious, it must continually generate major changes in the distribution of research facilities and resources inside and between different institutions. Such changes often require new administrative procedures, and must incur very heavy direct and indirect costs. Very large sums may have to be found for new instruments, in new buildings, on new sites, and equally large sums written off for obsolescent facilities and moribund research units that have not lived out their intended lives. Even when the accountants can prove that the final balance is strongly favourable, it is not so easy for hard pressed universities, for example, to find the capital required for major investments in future research capacity. The actual outcome, in terms of the 'health' of science, the vitality of institutions, etc., must depend on the sensitivity of the selection process, and the amount of money that is made available to facilitate it.

Even the most excellent of centres of excellence can become an administrative and intellectual dinosaur, isolated from the very needs it evolved to fulfil, and unable to adapt to new social or scientific demands. Rather than aggregating people and resources into large research entities in a few 'mega-universities', it may be more advantageous to create multidisciplinary regional research centres where teachers from various institutions can come and work on a day-to-day basis. Enhanced facilities for remote work [§3.6] may be less costly and more adaptable to changing scientific needs than new buildings and managerial structures.

6.9 Teaching and research

Whether or not a strong policy of selectivity will sharpen the top of the academic pyramid, it is bound to exclude original research from most institutions in the basal strata of higher education. There are already many institutions where undergraduates are taken to first degree level by teachers who have neither the facilities nor the time for more than token research activities. From now on, even the best departments in the better smaller universities may find it more difficult to compete nationally for resources, and may be forced to drop out of research altogether.

The partial separation of undergraduate teaching from direct contact with research seems to be an inevitable consequence of the transition to 'steady state' science. There is very wide agreement that the educational mission of a university benefits greatly if its teaching staff are active in research. These benefits, and the reciprocal benefits to research itself, are generally thought to outweigh the organizational complexities and professional ambivalences of combining the two functions. Nevertheless, even in a very rich country it would require a disproportionate national investment in science to make this possible for all institutions of higher education. The facilities required nowadays for serious laboratory research are just too expensive to be provided for everybody who is formally qualified to use them.There is no escape from a situation where only a minority of all university teachers will be in a position to do research and supervise postgraduate work, while the remainder will be expected to put all their efforts into undergraduate teaching and other activities that could not be described as advanced research [§7.14].

This is obviously a very significant aspect of the transition to a new regime in science. For some 200 years, the central tradition of academic activity has been the dual-function university, deeply involved in both transmitting and creating scientific knowledge. It is true that there have always been variants on this tradition. Many countries do not follow the British and American practice, where the universities that are most esteemed by undergraduate students are those where the active teaching staff are also in the top ranks for their research. Even in the UK, the polytechnics probably did a better job of teaching the less talented students than most universities, and throughout the United States there are outstanding 'liberal arts colleges' where the faculty do very little research. But there is a universal agreement on the supreme value of a professoriate whose members are distinguished both for their original contributions to knowledge and for their skill in communicating what they know.

Nevertheless, the advantages of this dual function are very intangible, and are not easily stripped of their anecdotal or ideological trappings. Indeed, administrative rationality, organizational flexibility and professional predilection all seem to point towards the complete separation of academic teaching from academic research. One cannot seriously fault the contributions to basic science of the laboratories, institutes or research units of the French CNRS, or the German Max Planck Gesellschaft, or even the British Medical Research Council, whose staff are highly professional research scientists with no direct teaching responsibilities. Why not simply follow the rationale of selectivity to its logical conclusion and hand over the science base to a system of full-time research institutes?

The standard counter-argument is that the bread of teaching needs to be leavened with the intellectual yeast of research. The converse may be just as true. In academic science, researchers need the mental challenges and alternative career opportunities offered by high-quality educational employment. Strong administrative measures are often required to save full-time researchers from going stale [§7.14] and to force specialized research units to adapt to scientific, technological and economic change. A loosely structured research entity embedded in an institution of higher education can do just as good work as a self-contained institute, with much greater personal and organizational flexibility. This may well be the best way for things to go as 'steady state' conditions take effect. But these are issues where individual careers should be given precedence over institutional structures.

7
Scientific careers

If I had a donkey that wouldn't go,
Would I beat him? Oh, no, no!
I'd put him in the barn and give him some corn,
The best little donkey that ever was born.

7.1 The demography of the scientific profession

The transition to steady state conditions is having a profound effect on research careers. Ever since science became a regular profession in the late nineteenth century, it has continually expanded in numbers and in employment opportunities [§4.1]. This has kept it a buoyant open-ended enterprise, where talented newcomers were welcome, and where they could look forward to opportunities for personal advancement right through their working lives. Yes, it was an uncommon profession, open only to a gifted, dedicated, minority. Yes, it was highly competitive, requiring exceptional tenacity to get to the top. Yes, it was not very well paid, and the reward of success was fame rather than fortune. But, even for the socially unambitious, it provided a secure, well-respected niche from which to explore nature, and seek honestly after truth.

Of course, the scientific profession has seen some hard times. At various periods, in various countries, military defeat, economic depression, or political repression have temporarily disrupted research careers. The biographies of many of the greatest European scientists of the twentieth century are punctuated with episodes of personal insecurity, hardship or exile: the stories of many of the lesser figures are even more tragic. In Mao's China, almost the whole scientific community was banished into squalid rural drudgery for a generation. The break-up of the Soviet Union threatens professional disaster to the members of one of the world's largest and most capable scientific communities. Generally speaking, however, modern science is the product of people living very secure, modestly comfortable lives, and provided with the time and the means to play an honourable part in an immensely successful and highly esteemed communal enterprise.

Resource constraints that put a bound on the total number of scientific jobs not only limit the openings for recruitment and advancement: they also affect the nature of those jobs and the attitudes of the people who fill them. The 'progressive' spirit of science thrives in an atmosphere of institutional growth and personal opportunity, where the established authorities do not have to be evicted to make way for new people, but can simply be left behind as new research areas are opened up for exploitation. The social organization of science is adapted to the ever-expanding 'Endless Frontier' of Vannevar Bush's famous post-War report. The expansion rate cannot be throttled back suddenly without severe structural strain.

To take an obvious example, the demographic effect of curbing rapid exponential growth is to increase the average age of the profession. Assume for simplicity that people are recruited at 25 and stay in science until they retire at 65. When the total number of research posts was doubling every fifteen years, the average age of researchers was about 35. Slow down to a 'steady state' situation, where new posts only become available as they are vacated by retirement, and this average must rise by something like ten years, to around 45. The steepness of the rise will depend, of course, on the rate of change in recruitment. If this is effectively discontinuous, as in the case of the tenured staff of UK universities in the late 1970s, this ageing effect will be exaggerated for a while as a transient 'bulge' passes through the age distribution.

This calculation is, of course, much too crude. The actual average age of academic scientists in the UK has certainly increased substantially during the 1980s, but not by as much as this simple model would suggest. The magnitude of the transient has been reduced by the voluntary, or involuntary, premature retirement of a number of older people, while the lower end of the age distribution has been sustained by recruiting relatively large numbers of younger research workers on short-term appointments [§7.9]. In other words, the approach to the steady state has not only darkened the career expectations of scientists; it has also had a significant effect on the social structure of the profession.

These structural changes are probably irreversible, at least as long as 'steady state' conditions prevail. Suppose, for the sake of argument, that an attempt were made to return to the traditional academic arrangements whereby most active scientists were recruited relatively early into a permanent life-long career of research. The replacement rate would then have to be far below the present rate at which researchers are now being trained. If recruitment to research training were then reduced in proportion, the proportion of graduate students and post-doctoral assistants in the scientific community would fall very significantly. The work of these 'apprentices' [§7.12] would then have to be done either by more senior scientists or by an enlarged and better-qualified corps of career technicians.

The actual *quality* of research would almost certainly suffer from this reduction in the proportion of younger people. The scientific enterprise is thought to depend enormously on youthful imagination and intellectual vitality. Although the evidence for this opinion is almost entirely anecdotal, there can be little doubt that an ageing community would be less adventurous in 'pushing back the frontiers of knowledge'.

There could be other, more subtle, effects. For example, scientists often judge their success by the influence they exert through the supervision of research students and assistants. This is the means by which they extend the range of their research, create 'schools' and pass on techniques, conceptual schemes, tacit knowledge and other research traditions. A significant motivating factor would be lost if scientists, on the average, could not expect to have more 'apprentices' during their whole career than the one or two needed to replace themselves.

Perceived limitations on recruitment into research training and employment would be transmitted downwards into undergraduate and school science education. Although society in general needs more technically trained people than ever before, students are already becoming disenchanted with 'irrelevant' and 'academic' research as a career and many university science departments are struggling to survive because of a lack of adequately qualified applicants. If 'steady state' science were seen as a closed profession with poor promotion prospects and low salaries, it would no longer offer an attractive career to able young people.

7.2 Destabilizing a secure career

'Steady state' conditions have dislocated the traditional career patterns and personnel practices of academic science beyond repair [§5.11, §6.4]. Return to a system of early recruitment to secure, life-long employment now seems unlikely – perhaps even undesirable. In countries such as France, where the tradition of academic tenure is still strongly entrenched ideologically and institutionally, it creates more and more organizational and intellectual difficulties. Rather than limiting initial recruitment into the research profession, many countries seem to be moving towards greatly increased personal competition for research employment throughout each career, accompanied by substantial attrition at every stage.

Some great universities, such as Harvard, have long practised a conscious policy of 'up or out' for their academic staff. All low-status posts were temporary, and could be held only up to a certain age: those who failed to get promotion on to the 'tenure track' had to seek their fortunes elsewhere. An extension of this policy to cover all research posts, with the possibility of 'relegation' to another type of job at any point in a personal career, would be consistent with other features of 'steady state' science, such as increased institutional differentiation and competition for resources [§6.8]. An economist might argue that the opportunity to do research – that is, access to expensive scientific facilities – has become a scarce commodity, to be made available only to those who can show that they can make the best use of it.

This tough-minded policy obviously has its attractions for those who believe that the science base needs to be managed more systematically and efficiently, in order to keep up with economic and technological change. From this point of view, academic tenure [§7.9] is seen as a major barrier to managerial initiative and personal upward mobility. By ejecting, or redeploying into teaching, the substantial numbers of older scientists who have not apparently lived up to their promise, institutions can keep down the average age of their active research staff, and motivate people at all levels to be energetic and productive.

The conditions for a more open and competitive career pattern in research are already present in the UK – witness the lack of jobs for PhDs to match their specialties, the growing proportion of mature scientists on short-term research appointments, and reduced opportunities to enter effectively permanent research employment [§7.13]. But this transition is still far from complete. Although much weakened legally, customary academic tenure is still widely honoured, in that university staff in established posts are very seldom made redundant. Indeed, a two-tiered social structure has developed, where *de facto* tenure is not, as it used to be, primarily associated with age and experience, and where many competent scientists are still moving through a succession of temporary posts in mid-career.

What we are seeing here is a very significant decline of *security* and *stability* in research careers. Later in this chapter we shall consider the cultural and epistemological effects of the loss of what have long been such characteristic features of academic life – effects that could be very serious. But even in the hardheaded terminology of personnel management, career insecurity and instability are not necessarily advantageous. For example, they do not make science an attractive profession for prospective entrants. They may combine with the inherent uncertainty of research to produce continuous anxiety, so that perfectly competent scientists become excessively cautious in their research proposals or are even driven out of research altogether.

In the past, research careers were highly valued. It is true that apprentices and aspirants often went through a long period of genteel poverty before they became fully established and that even the top people in academic science were not paid so handsomely as people of comparable ability or social standing in other walks of life. But the salaries of university professors in most Western countries have always been adequate for their style of life and professional needs, and were associated with very attractive benefits of high public esteem, security of tenure, and almost complete personal autonomy.

Present policies are eroding these enjoyable perquisites of university life. In the UK, even academic salaries have fallen behind those in other comparable professions. And yet it is clear that additional material rewards will be needed to compensate for the relative loss of public esteem, job security and personal independence. In principle, a free salaries market should develop, with competition between institutions for the ablest or most promising researchers. In practice, chronic shortage of funds for research is likely to keep salary costs down, with the inevitable consequence of a 'brain drain' out of academia [§7.13] or out of any research system that is not able or willing to pay its researchers at internationally competitive rates [§8.10].

7.3 The competitive tradition in science

Science has always been an intensely competitive calling. Scientists have always vied with one another, often quite openly, for academic posts and public esteem [§6.7]. Their research claims have always been subject to communal criticism, very often to the point of devastating destruction. 'First past the post' is the rule in scientific discovery, and there are no consolation prizes for those who also ran. The story of many outstanding achievements in science is not of lonely heroism: it is of a scarcely concealed race towards a half-perceived goal. *The double helix* may not be an objective account of how Francis Crick and Jim

Watson discovered the structure of DNA, but it captures vividly the spirit in which so much good science is done.

Personal rivalry is not just one of the deplorable aspects of the scientific endeavour: for the serious players, it is the name of the game. Selective pressure at the level of the individual researcher and his or her contributions to science is the key – the only key – to excellent performance and reliable results. Philosophically speaking, the most convincing general theory of scientific knowledge is an 'evolutionary epistemology', where the validity of a research claim is tested in a struggle for survival with competing claims. Sociologically speaking, this is associated with a contest between the champions of these various claims – a contest that is fuelled by such public rewards as academic posts and honorific prizes.

Great scientific discoveries win great prizes and great fame. The razzamatazz surrounding the annual Nobel Prize awards clothes each prizewinner with a personal aura, on into eternity. But this is not just a spectacle arranged for the benefit of a few star performers. There are also smaller prizes, and humbler halls of fame, for the smaller achievements of lesser mortals. This applies right down to the lowest levels of the research world. The essence of the academic tradition is that scientists are appointed and promoted, often in direct competition with other applicants, on the basis of their public contributions to knowledge.

This is a major factor in the organization of research. For example, one of the reasons for the internal fragmentation of universities [§6.7] is that many of their most effective academic staff do not nail their flags to their institutional masts, but see themselves as potential competitors for the most prestigious jobs in their field, wherever these may happen to arise. 'So old Bloggs is retiring at last from the Camford chair. I wonder if they will advertise it? How are my chances? Is Coggins going to get it? Well, if they invite me to apply, I'll certainly put my name in. These last ten years at Ambridge have been very good, and they have treated me very well, but I owe it to my career to go to the best place. After all, I could probably take my research group with me . . . '

In effect, many of the most able scientists in a particular field continually watch their stalls in the national or international *job market*, ready to strike a good bargain if the opportunity occurs [§8.10]. Each is a vendor of a highly prized commodity: his or her own research promise and grant-earning capacity. Universities and research institutions, at home and abroad are ready customers for the highest-quality products in this market, seeking out what they judge to be the best and offering a very good price in terms of a personal stipend and research facilities.

This is not a novel analogy. In *The academic marketplace*, first published in 1958, Theodore Caplow and Reece McGee described vividly how the academic profession in the USA hinges on the elaborate processes by which universities recruit their staff. Indeed, these processes have become so complex and institutionalized that other social metaphors suggest themselves – the 'slave market' of post-docs at the annual subject convention, the 'patronage' dispensed by the 'godfather' of the discipline, the 'head hunting' for 'distinguished professors' to adorn the faculty, and so on.

In fact, this form of competition between universities dates back to at least the early nineteenth century, when it became customary for German universities to bid against one another for the most promising scholars, and where it was accepted that moves from university to university up the hierarchy of esteem were normal steps in a scholarly career. The same tradition of inter-institutional mobility became established in the UK, although never with such intensity as in the USA.

The driving force in this competition is, of course, the dependence of institutional performance on the quality of its academic staff. Research is extremely labour intensive, yet research excellence is quite rare, and not always easy to spot. If an institution is to be successful in obtaining research funds, it must have researchers able to hold their own in the project markets [§5.11, §6.8] – and these are always in short supply. If, as was usual until quite recently, they are likely to gain early tenure [§7.2], they have to be selected with all the care that a manufacturer gives to the purchase of a piece of capital equipment, such as a machine tool, which has to perform well for another 30 years.

Competition in the academic job market thus plays a very important part in the larger struggle for resources between institutions. In these hard times, it is scarcely surprising that this competition bears more heavily in individuals, and extends to the later stages of their careers. Many significant features of contemporary academic life can be interpreted as naive efforts to make this whole process more economically efficient. The market is made more 'perfect' by removing contractual constraints such as job tenure, strengthening bargaining power on the demand side by having an excess of researchers on short-term contracts, improving process control by separate accountancy of teaching and research [§6.9], and eliminating obsolete items by career appraisal and performance evaluation. But whether the invisible hand works optimally here is another question, which we must consider more deeply.

7.4 CUDOS or PLACE?

Scientists are important as people, not just as *personnel* or as competing *entrepreneurs*. There is a large but fragmentary literature about the problems that arise in the personal careers of research workers in universities and other institutions devoted primarily to basic science, but most of this literature is directed at particular issues as they arise in particular countries. Vice-Chancellors, Rectors, Deans, Departmental Chairmen, Directors of Establishments, Heads of Laboratories and other responsible leaders often acquire great expertise in handling these problems, but this experience is not systematically recorded or analysed.

This is not because the management of their human resources is unimportant to academic institutions. On the contrary, the million or so people around the world who are affected by it talk about it constantly. But there is a strong ideological inhibition against referring to it in those terms. 'We scientists are individuals', they say, 'not resources to be managed!' This *cri du coeur* encapsulates the whole subject.

To make sense of this situation, it must be analysed a little more deeply and clearly framed. This is not so easy, because 'academic research' is often treated as just a small part of a system that includes all the applied research and technological development undertaken by governmental and commercial organizations. These organizations have missions which require the systematic deployment of all their available resources, often in large teams directed urgently towards specific goals.

We have already referred [§6.6] to the widely documented art of 'R&D management' which has evolved in these circumstances and which forms a natural sub-category of management studies in general. The career and personnel problems of qualified scientists and engineers (QSEs) working in R&D organizations [§7.13] are not so very different in principle from those that arise in other large bureaucratic organisations, and are usually dealt with along the same lines. The evidence is that people initially trained in academic research [§7.12] – for example, with PhDs or even some post-doctoral experience – adapt readily enough to this working environment, and get adequate satisfaction from producing what is required of them by their managers.

Nevertheless, as we have seen in the previous chapter, academic science fails lamentably to conform to the general principles of R&D management. This is not to say that it should be treated as a traditional way of life, to be subsidized and sheltered on its tribal lands for the sake of its beneficial natural products. Technical progress necessitates less individualistic and much more expensive working methods. Societal patronage is not unlimited, and is accompanied by increasing demands for accountability and utility [§4.2]. The general process of collectivization inevitably pushes academic research towards the cultural practices of conventional R&D organizations.

What is surprisingly unchanged, however, is that academic researchers are still apt to describe their work in *ideological* terms. They do not say 'I'm just doing this as a job': they insist that they are responding to nobler aspirations, such as the search for truth, the pursuit of knowledge for its own sake, the satisfaction of curiosity, etc. In other words, they refer themselves to an academic *ethos*, supposedly governing the scientific life.

At this point, a little formal sociology can be extremely helpful. In 1942, Robert Merton showed how this ethos could be analysed into a set of quite general *norms*, which tacitly govern many of the traditional practices of academic science. Although this type of functional interpretation is now considered very questionable among sociologists of science, the Mertonian scheme still provides a very convenient conceptual framework for an account of the changes that are taking place in scientific roles and careers.

In particular, this analysis throws into sharp relief the peculiarities of academic research within the larger R&D system. A research scientist who is trying to conform to the Mertonian norms is not very amenable to the prescriptions of conventional R&D management, which are based on quite different principles. In this light, 'human resource management' in academic research shows up as a very different game, which needs to be understood in its own terms.

Very succinctly, the Mertonian norms are as follows:
- *Communalism*: Science is a collective enterprise, and the results of research should be made public at the earliest opportunity.
- *Universalism*: Participation in the scientific enterprise should be open to all competent persons, regardless of race, religion, nationality or other affiliations.
- *Disinterestedness*: Scientists should present their results impartially, as if they had no personal interest in their acceptability.
- *Originality*: Research claims must be novel; copies of previously published work are not acceptable.
- *Scepticism*: All research claims should be subjected to critical scrutiny and testing.

These norms are obviously very schematic and idealized. They set standards which very few academic researchers have ever achieved in practice. They are also incomplete, in that they say nothing about the philosophical principles on which 'originality' or 'scepticism' should be based, and they give no indication of the nature of the institutions where these norms might be practised. This is essentially an ethos for scientists as *individuals*, seeking the personal '*CUDOS*' spelt out by the initial letters of these norms.

The *CUDOS* scheme thus assumes a social framework in which the individual academic researcher can pursue a *reputational* career. Public contributions to knowledge are rewarded by communal recognition in the form of citations, honorific titles, prizes and, above all, academic employment. Work hard and long for your PhD, and your seniors may find it interesting enough to give you a post-doctoral fellowship. Publish a few papers that are cited favourably by other scientists, and they might put you on the tenure track. Acquire a name as an international authority on your subject, and they ought to make you a professor. Make a famous scientific discovery, and they will beg you to accept a highly paid appointment at an elite university. Employment that allows time and resources for research is, of course, absolutely essential for such a career, and is often hard won. But the other responsibilities of this employment, such as lecturing, examining, administration, etc. tend to be regarded as incidental to winning further esteem as a researcher.

In other words, this scheme defines and celebrates an *individualism* that is clearly inconsistent with the *corporate* spirit of non-academic R&D. The basic assumption there is that people are pursuing *organizational careers*, where they make their way up a specific enterprise by directly contributing to its mission. Labour long to produce reliable test data, and Techsystems Ltd will promote you from Assistant Researcher to Researcher. Suggest a solution to a design problem, and Techsystems Ltd will make you a Group Leader. Manage the project development team effectively, and you could be in the running for the job of R&D Director. Turn a bright scientific idea into a highly profitable product, and you might even end up as Chairman of Techsytems Ltd, with loads of responsibility and money.

The principles governing such careers are so much taken for granted that they are seldom made explicit. My own suggestion is that they can be summed up in the acronym *PLACE*, indicating the nature of the reward that can be won by performing *P*roprietary, *L*ocal, *A*uthoritarian, *C*ommissioned and *E*xpert work. These principles are so straightforward that it scarcely seems necessary to define them further. Like the Mertonian norms, they are very simple and schematic. But they encapsulate many features of employment as a research scientist in a typical R&D organization such as an industrial laboratory or government scientific establishment.

It is quite obvious, moreover, that the corporatism underlying these organizational principles is almost impossible to reconcile with the academic ethos. Indeed, many career problems now arising in academic research can readily be interpreted in terms of the practical contradictions that arise between the tacit demands of *CUDOS* and the more explicit principles of *PLACE*. In some circles it is supposed that the latter scheme is completely superseding the former, in an inexorable process of rationalization and modernization. Whether or not this is so, many of the personal and personnel issues of 'steady state' science can best be understood in terms of the conflict between these two cultural forms.

7.5 World markets for research claims

The very first of the *CUDOS* norms, that is 'communalism', lays great stress on the *publication* of research results [§2.7]. In the everyday language of academic science, a researcher is said to make 'contributions' to knowledge – that is, books and articles in scientific journals reporting experimental measurements, observational data, theoretical calculations, speculative hypotheses, instrumental techniques, critical surveys, reinterpretations of old observations, and so on. These texts are extraordinarily diverse in style and content, but they have two features in common: they are open to the world, and they make scientific claims.

These two features are closely linked. Scientifically speaking, the significance of a research claim depends on how it stands up to public scrutiny and comment by the relevant scientific community. However astonishing a research result may seem, it does not officially count as a potentially important scientific discovery until it has been disclosed to the research world, and it confers little credit on its discoverer until it has been more or less accepted as valid. However esteemed a scientist may be, the soundness of his or her work is not taken for granted. It is refereed before publication, subjected to expert questioning at seminars and conferences, reviewed critically in survey articles, cited favourably or unfavourably in other papers by other scientists, and tested by replication or experimental refutation.

This is not just a quaint tribal custom: it is the social mechanism by which reliable scientific knowledge is generated and evolves. In effect, the literature of a research specialty is a public marketplace, where the pigs and potatoes produced by peasant farmers are prodded and weighed and shrewdly purchased by other peasants. A successful reputational career can only be achieved by, so to speak, setting up a stall and displaying one's wares in this market. Underlying the academic ethos is a non-contractual but well-established social process whereby the scientific community receives 'contributions' and provides 'recognition' in exchange for those it values.

This notion of communal recognition is very subtle and delicate, but nobody who has observed the ferocity with which scientists vie with one another for it will doubt its psychological reality. That is why it is essential to have a convention that a research finding is 'owned' by the person who first reported and claimed it publicly. Without such a convention, the whole social mechanism that generates relatively reliable scientific knowledge would fall to pieces.

Evolutionary accounts of scientific and technological progress are built upon the efficient working of this mechanism. As Darwin perceived, the market is prodigal with goods, but the invisible hand of selection is precise. International competition for the validation of research claims is very harsh and very discriminating. A hundred flowers bloom for the one that is finally selected for seed. One paper with a hundred favourable citations is worth infinitely more than a hundred papers with one citation each.

This selective process is strengthened by the norm of 'universalism'. What we might call the research claims market is a *worldwide* social institution, beyond the control of any one national scientific community [§8.1]. In practice, this is sub-divided into a complex system of imperfectly coordinated specialty markets, corresponding to the familiar hierarchy of academic disciplines and research areas [§8.2]. But each of these specialty markets is more global, more formalized, more objective and less vulnerable to manipulation than the reputational markets with which it is linked. It is not driven by some abstract striving for truth. It is driven by fierce competition between individuals desperately seeking public certification of their claims.

This is a highly idealized rationale of the system of customary practices and habitual procedures which has grown up around academic research. The abundant evidence that this mechanism is very imperfect leads many sociologists of science to dismiss it out of hand, in favour of models where a hidden play of social influence and power is revealed. Nevertheless, in its ramshackle way, this is how it has worked in the past, and how it still governs the lives of a great many people. That is why access to, and active participation in, the relevant specialist submarkets of world science is considered a fundamental condition for a career in academic research. It is not just a matter of maintaining efficiency in terms of competitive excellence in performance. The intellectual integrity of academic science in any country depends on putting the findings of its researchers on public display internationally, and accepting fully the value put on these findings by the world community.

7.6 Intellectual property

The knowledge that comes from research can not only be traded for 'recognition': it can also be sold for hard cash. Basic science nowadays is seldom totally detached from its potential applications. Commercial firms are continually on the lookout for academic research findings that might be exploited technologically. From this point of view, all new scientific information is potentially 'intellectual property' [§2.7], with a legal owner empowered to demand payment for its use.

Outside academia, this is a very familiar concept. The whole job of the scientists working in an industrial laboratory is to create proprietary knowledge – that is, to obtain research results that can be turned to profit by the firm that employs them. This is what they are hired to do, and the fruits of their labour belong to their employer. The same applies in principle to most public sector R&D organizations, where researchers are usually required to obtain formal permission to publish any of their findings.

The very idea that the results of research should be considered 'proprietary' offends against the academic norm of 'communalism'. But this principle is the cornerstone of any scientific career directed towards gaining an organizational '*PLACE*' rather than personal '*CUDOS*'. Indeed, the most lucrative technical knowledge is often distinctly 'local' (for example, the recipe for manufacturing a commercially successful explosive) rather than 'universal' (for example, the theory of highly exothermal chemical reactions) and the sphere of action of an organizational career as a scientist is correspondingly bounded.

In spite of all the rhetoric acclaiming scientific knowledge as a public good, the majority of research scientists are actually employed to produce results that can be turned to private benefit. The important role of commercial secrecy, patent law, technology transfer agreements, etc. in the process of technological innovation is not our subject here. The interesting point is that the literature on personnel management has little to say about the disposition of intellectual property rights inside R&D organizations, except where these might affect job mobility outside the firm. In general, most of the people working in such organizations seem quite content to follow compartmentalized scientific careers.

The transformation of scientific information into a marketable commodity is affecting the structure of science as a vocation. For an academic scientist, the jealously guarded right to have one's name publicly attached to a scientific discovery is not just vanity. It represents a genuine form of intellectual property, which may eventually be of considerable material value to its owner. Although this right may sound somewhat nebulous, it may not be at all advantageous to exchange it for, say, the uncertain financial return on a possible patent. A situation is developing where a university researcher may even lose any personal stake in the property he or she has produced. It is now becoming a serious practical question whether the right to exploit the results of a project should be vested in the organization that has funded it, in the institution where the work was done, or in the individual researcher who probably thought up the research problem in the first place?

The co-existence within 'science' of communities following two such contrary career patterns caused no grave difficulties provided that they were carefully segregated. Even now, some academic scientists still insist on the absolutely fundamental distinction between the 'pure' research that they do, without the least concern about its possible eventual use, and the 'applied' science that goes on in other places. But, as the concept of 'strategic' research shows [§2.5], scientific knowledge can no longer be classified like that.

The institutional distinction is also impossible to maintain under 'steady state' conditions, where every organization putting money into research is under pressure to get what cash return it can out of its investment. Universities enjoin their academic staff to look for every opportunity to exploit commercially the results of their research, and especially to protect all intellectual property rights (IPRs) that might prove to be of value. This applies particularly to research contracts with private sector firms, which normally have a very active interest in the applicability of the work they commission [§6.7]. Universities thus find themselves making different intellectual property agreements with different classes of customers. One might then question the role of public sector customers for basic research. Are they buying a global public good, an appropriable competitive advantage for a home-based firm, or a lottery ticket for a fortunate institution?

Now it has to be said that the market in IPRs is highly speculative. There is an immense variation in the amount that can ultimately be realized commercially out of a particular academic research finding, even when thoroughly protected by patents. In fact, 99 per cent of what research scientists discover is absolutely worthless in commercial terms. The total value of the merchandise is large, but, like natural pearls in oysters, most of it is concentrated in only a tiny proportion of the goods on offer. University authorities are attracted to this type of speculative enterprise by stories of occasional spectacular windfalls – especially the lost opportunity of the Medical Research Council to cash in on the enormously valuable invention of monoclonal antibodies. The evidence is that the average return to universities on patents arising from basic research scarcely covers the cost of filing them.

Academic researchers themselves are not averse to putting their research claims on to the commercial market in the hope of making a killing. But this competes uncomfortably with the reputational market, where the same research claims may turn out to have quite a different type of value. Unfortunately, these alternative valuations are incommensurable, since the commercial value often depends on temporarily keeping secret what only has reputational value if made public immediately. The two markets also have quite different conventions concerning the 'ownership' of the intellectual property in question. It is notorious, for example, that when a team of academic scientists makes a significant discovery, the public acclaim is not shared out among its members, but goes almost exclusively to its leader. In a well-run industrial R&D organization, by contrast, great care is taken by the management to apportion the rewards for success fairly among all those who have contributed to it.

There is a deeper problem. The exchange of 'recognition' for 'contributions to knowledge' is not a normal market transaction. Indeed, there is a strict taboo against any 'trading' between those who seek esteem and those who award it. The proper analogy is not with monetary commerce or even barter, but with a traditional system of potlatch. At periodic intervals, the tribe indulges in an orgy of public gift-giving in which those who give away the most valuable goods receive the greatest honour. In other words, researchers are expected to present their work as if they were, indeed, just 'honest seekers after truth', with complete disdain for any suggestion that they might be getting something in return for their gifts to mankind. The psycho-dynamics of such practices is better understood through social anthropology than through economics.

The academic norm of 'disinterestedness' not only applies to individuals: it also keeps the interests of scientific institutions at arm's length. No matter how much a university may hope to benefit eventually from a valid scientific discovery by one of its staff, it must keep out of the validation process. As was clearly shown by the furore over the behaviour of the University of Utah in the 'cold fusion' affair, any attempt by an institution to assert 'ownership' of a publicly communicated research finding is bound to discredit that finding. It is only because the IPRs of academic researchers are tacit, personal and not negotiable that their public claims can be assessed reasonably objectively by the relevant research communities. This delicate mechanism would be destroyed if it were not firmly protected by convention from the corporate interests of large institutions.

In reality, universities have a very large stake in the reputational success of their academic staff [§7.3]. For a world-famous institution such as Oxford or MIT, the money that comes to it in the form of grants and gifts because of the scientific eminence of its professors must enormously exceed the income that it earns from the actual sale of the intellectual property they generate. Indeed, the university funding councils are trying to allocate resources explicitly on just these lines. Their numerical research performance indicators [§6.1] are calculated according to the criteria that might be used, say, by a sectional committee of the Royal Society assessing a candidate for a Fellowship – for example, on the basis of articles published in refereed journals. The senior academics who run the block grant system are so steeped in the conventions of the individual reputational market that they try to apply them directly to whole departments and institutions.

7.7 From patronage to career management

In spite of its uncompromising individualism, the academic ethos is surprisingly reticent about the concept of a 'career'. In fact, the mysterious norm of 'disinterestedness', which seems to contradict the blatant devotion of every scientist to the furtherance of his or her own scientific ideas, can be interpreted as a health warning against the seductions of 'careerism'. All that the successful researcher is promised is more and more *CUDOS*, right to the end. The scientific life is idealized as a *path* through an unexplored wilderness, not a *ladder* leading up to commanding heights.

Many scientists and scholars do, indeed, seem to get all the satisfactions they need in life from doing research and receiving the customary tokens of communal recognition for their achievements. Sometimes they move opportunistically from problem to problem, without any conscious plan; sometimes they set themselves very distant, apparently unattainable goals, and devote themselves heroically to struggling towards them. When successful they are obviously gratified by the knowledge that their efforts were considered eminently worth while by their peers, but they emphatically deny that the thought of fame was the spur that drove them on.

It has always been the duty of the world of science, through its various institutions, to foster this way of life, to provide the facilities to pursue it where these are obviously merited and to ensure that at least a few outstanding and persistently creative researchers are appropriately rewarded, materially and honorifically. It may be that modern science goes to excess in the scale of its public rewards to a limited number of pre-eminent individuals. There is no need here to go into the element of 'undue esteem' that can be so very damaging to some academic research careers – the hype surrounding Nobel Prizes [§7.3], the consuming jealousy of success, the fraudulent discovery claims, and so on. Such phenomena are not really new, although the expanded public image of science has made them more notorious. One might say that such pathologies are inevitable in any social system where reputation plays such an important role.

The key point is that universities treat their established academic staff almost as if they were self-employed in this aspect of their working lives [§6.7]. They are praised for being 'self-winding', they are encouraged to win support for their personal research projects, and their intellectual freedom is protected by tenure. Academics customarily refer to research as 'doing their own work', as distinct from the lecturing and other work that they do for their institution. One might say that every academic post has attached to it the life tenancy of a small farm, where the holder may spend a good part of his or her time labouring to make a personal living.

Paradoxically, the framework that supports this liberal tradition may be quite rigid and formal. Although UK universities are legally independent corporate bodies, they were for a long time subject to central financial controls, which inhibited them from competing with one another for academic staff. Until recently, grades of seniority and pay scales were negotiated and standardized across the system, and tenure and pension rights accumulated in one institution were automatically transferred if one moved elsewhere. In effect, academic employment in a British university constituted membership of a single elite corps, almost equivalent to the formal civil service status of academics in countries such as France. In the same spirit, members of the UK scientific civil service doing quasi-academic research were encouraged to feel almost as free as university researchers in their conditions of employment.

This concept of a highly personal reputational career shields academic scientists from institutional interference in their research plans [§7.6]. It also inhibits institutions from being actively concerned with the professional careers of their employees. This applies at every level. Success in the reputational market is, of course, applauded. The university goes out of its way to congratulate its professors when they are elected to an honorific academy or receive a public honour. But non-success is passed over in embarrassed silence. As we have seen [§6.7], conventional managerial attitudes to personnel problems are strongly resisted, even though these could help the substantial proportion of academic researchers who prove to be inadequately equipped for the strenuous competition of a world-wide academic community.

The studied indifference of universities to the fate of their academic staff is most noticeable in the case of 'contract researchers' – that is, academically qualified researchers employed on short-term contracts. The tacit assumption is that they take on such temporary work voluntarily, to gain experience and recognition, and that they will either win their way to an established post or fall by the wayside into more gainful occupation. Academic institutions and their senior managers do not feel that they have the same continuing responsibility for them as they would if they were seen to be following organizational careers under *PLACE* conditions.

One of the reasons for this indifference is that young scientists were usually very dependent on personal *patronage* in their initial stages [§7.12]. The trials and tribulations of the academic apprentice will be familiar to anyone who has read the memoirs of that great cohort of scientists who came to maturity in Europe between the World Wars. The head of a large research school, producing many PhDs, was a powerful figure. Yes, the star pupils of the star professors were well looked after, and came early into their own. But the work of many talented young researchers was often exploited by their less talented seniors, to whom they might have to defer for years in order to be recommended at last for an independent post. Success in the academic job market is, of course, still heavily influenced by such factors.

This was not necessarily a sinister or one-sided relationship. At higher levels of esteem, the patron's own reputation could depend in the long run on the public achievements of the more junior scientists whose work he or she was known to have advised, encouraged and supported through adversity. One of the ways of acquiring *CUDOS* is to recognize early on the unsuspected capabilities of an unconventional researcher: one of the ways of publicly displaying scientific mediocrity is to fail to promote the careers of manifestly able students or junior colleagues. A significant defect of the peer review system [§5.4] is that the benefits of patronage are not reciprocal. An anonymous panel of specialists awarding a research grant to a speculative project cannot expect to gain credit for a decision that might turn out to have furthered a successful scientific career.

7.8 'Doing my own work'

Although academic patronage is often very coercive, it is not strictly *authoritarian*. Beyond the stage of formal tutelage, the patron is not empowered to issue instructions on what research should be done, or how it should be carried out. It is thus quite different in principle from the managerial relationships that shape the careers of scientists in industrial firms and other R&D organizations [§6.6].

To say that industrial research is 'commissioned' may exaggerate the extent to which this authority is used in practice. The art of R&D management is to get scientists and other professional employees to devote themselves personally to the goals of the organization, and to carry forward its general scientific and technological policies precisely as if they were their own. Quiet corners can be provided for those who work best undisturbed, while there is tolerance of the highly innovative entrepreneur as a product champion within the organization. But even in the most permissive industrial or government laboratory, researchers have to obtain periodic approval for their research objectives and work programmes. They knowingly forego the right to choose their own research problems in order to qualify for a *PLACE* in the organization. By giving up this freedom, they put upon the firm or government agency the ultimate responsibility for ensuring rightful progress in their professional careers.

By contrast, the foundation stone of the academic ethos is the notion of *autonomy*: personal freedom to choose an area of research, decide what scientific problems to tackle in this area, and set about tackling them in one's own way. This is held to be an essential requirement for the 'self-winding', 'lonely seeker after truth', who is 'inspired by endless curiosity', to 'do her own work' and thereby contributes in the highest degree to the 'advancement of knowledge'. Such is the image invoked to justify a whole repertoire of academic practices centred around academic tenure and collegial self-governance.

Of course this is a totally unrealistic requirement derived from a highly idealized notion. Any serious account of the life and work of any successful scientist will bring to light many occasions when they were well aware that their research options were very narrowly constrained by their personal skills, material resources, instrumental facilities, institutional commitments, career goals and other very cogent considerations. The interesting thing is that these constraints are always systematically discounted, or treated as barriers heroically surmounted, in the name of scientific individualism. The quality of a scientific discovery is enhanced by the notion that it arose out of the exercise of this autonomy – for example, by fastening on an unfashionable problem, by pressing ahead with an apparently fruitless investigation, or by turning aside from other work to follow up a chance observation.

The fact that this notion of freedom is sometimes illusory does not make it undesirable. On the contrary, a great deal of 'personal space' is an essential operational requirement for many forms of scientific creativity. More to the present point, a tradition of personal autonomy is a psychological necessity for individuals who are expected to take full responsibility for the success or failure, the maturation or frustration, the triumphs or disasters, of their scientific careers. This can be a very heavy personal burden. Academics are well aware that research reputation is only won through exceptional individual effort and enterprise. In particular, it requires *originality* in the choice of problems and methods – an originality that is not always demonstrable through projects that maximize institutional profits.

This norm of *originality* is extraordinarily difficult to satisfy. Criticism is inexorable. Academic researchers have to strive for years to receive communal recognition for their contributions [§7.5]. In practice, the only obvious way to become internationally competitive is to *specialize* intensely in the field of one's research. Extremely narrow subject specialization [§3.9], often for a whole career, is not just a reprehensible academic tradition: it is the rational response of the individual researcher to the demands of the research claims marketplace. Only by getting to know a specific subject very thoroughly can a scientist be sure of choosing research problems that other scientists have not yet thought to attack, or have not acquired the skill to solve.

This has two contradictory implications for scientific careers. On the one hand, academic researchers actualize themselves through the acknowledged mastery of their respective research specialties. It means a great deal psychologically to be the world authority on some particular scholarly matter, even if this is very pedantic and insignificant in relation to all the rest of knowledge. But such a status does not last unless it is supported by the further contributions required to keep up with the moving frontier. It seems imperative to be free to go on working in 'one's own field' – with all that such freedom requires by way of research equipment, assistants, dispensation from other duties, etc.

On the other hand, academic specialization can lead to the pathological condition of 'undue persistence'. Individual researchers often become unproductive and stale in mid-career, through trying to stay too long in the same narrow field. This condition is not unknown in non-academic R&D organizations, but it is widespread in academic research, and probably wastes a great deal of research talent and experience. It cannot be dealt with at all in the *CUDOS* framework, where the responsibility for problem choice rests firmly with the individual researcher, obsessed with a specialty and becoming aware of his or her plight too late to make a change.

These latent tensions are clearly revealed under 'steady state' conditions. Extreme specialization obviously imposes very severe constraints on any top-down programme for, say, the national science base. Academic scientists naturally oppose any shift of priorities and resources requiring them to move from the fields in which they have invested so heavily. And yet they would benefit from a system of personnel management that would facilitate such moves. This is an issue to which we shall return later in this chapter [§7.14].

7.9 The challenge to autonomy

This tradition is now being challenged. Academic tenure, the ultimate bedrock of individual autonomy in research, is being phased out or trimmed down [§7.2]. The proportion of established university posts has been reduced, so that most fully qualified researchers have to wait for some years in a succession of temporary appointments before they get effectively permanent employment. Many remain in this condition all their working lives, taking on a series of short-term contracts to carry out specific projects which they have not themselves conceived, by methods which they have not themselves designed, to get results for which they will get no scientific credit. In effect, a great many academic scientists now have so little autonomy in their research that they can scarcely be said to be pursuing reputational careers in the traditional sense.

Tenure as such, in the sense of a well-paid research and/or teaching appointment, protected against every eventuality until the age of retirement, is not the real issue. The point is that permanency of employment, whether legally protected or merely customary, has always been the cornerstone of 'academic freedom'. It is the basis for many features of the scientific life, such as life-long commitment to a particular specialty, acquisition of expertise, building up a reputation as an authority, and personal autonomy in the choice of problems, which have always been considered essential for the exercise of scientific creativity.

The formal abolition of academic tenure in UK universities is thus only the visible sign of the erosion of many of these liberties. To some extent, this is a straightforward consequence of collectivization. As we have seen [§3.7], in most fields of science, researchers can no longer work effectively in isolation, but have to band together in larger and larger groups. It is one thing for a few independently minded and independently based scientists to collaborate voluntarily on a scientific project leading to a joint publication for which they may all claim equal *CUDOS*: it is quite another matter when dozens of qualified scientists have to work together for years as a single, highly organized *team* to produce a single paper with 57 'authors'. This condition, which has long been normal in industrial R&D, is now becoming common in certain fields of academic research.

The most extreme case is in high-energy particle physics. Go to CERN [§3.6, §8.5], for example, and you will find several hundred full professors, lecturers and graduate students from a score of universities, devoting all their scientific energies to the design and performance of a single enormous experiment. For the sake of courtesy, such enterprises are called 'collaborations', but the reality is that the great majority of those involved have surrendered their personal scientific autonomy to the collective authority of the team – usually directed by a very powerful leader. This may give them excellent opportunities to demonstrate their technical capabilities. An individual may be highly valued within the team as a wizard at some sophisticated technique, such as software design, or electronic engineering, or data analysis. But such a *local* reputation is not automatically disseminated in the general academic marketplace through the published account of the research.

For the moment, this type of highly organized teamwork rules the careers of only a minority of academic scientists. But the great majority are subject to more subtle pressures because of the significant resource constraints associated with 'steady state' conditions. Internal economies now dictate heavier teaching loads, more unevenly distributed among staff members, in most university departments [§6.9]. Even scientists who are well established, with permanent posts in highly regarded institutions, are forced to submit their research plans to others for scrutiny and approval. Not having personal control over the resources they need for their research, they have to seek support from 'patrons' or 'customers' who have their own notions of what should be done.

In the worst case, financial pressures may induce a competent academic scientist to enter into a contract for routine research which will add nothing of value to his or her *curriculum vitae*. A university pharmacological unit, for example, may be forced to make a living by testing new drugs for a pharmaceutical company, or a department of materials science may be heavily engaged in preparing new alloys for an engineering firm. The work that many academics do as industrial or commercial consultants, although often very well paid, is not of their own choosing, and seldom contributes to their scientific or scholarly standing.

In practice, as the experience of R&D organizations shows,

the majority of working scientists can adjust satisfactorily to a situation where they have limited 'strategic autonomy' in the choice of research problems [§7.4]. Many of them do not actually find it easy to set themselves new scientific goals, and are reasonably happy working towards the fixed, short-term objectives that are often essential in getting useful things done. What they resent is interference with their 'technical autonomy' – that is, the precise way that they reformulate 'problems' into 'projects', select their methods and instruments, design experiments, collect data, analyse their results and present their conclusions.

Many scientists are made quite anxious by uncertainties concerning the problem areas on which they should concentrate. They do not necessarily regard it as an infringement of academic freedom to have to give up personal control over the long-term goals of their research. Nevertheless, loss of strategic autonomy is generally reckoned to be one of the most objectionable features of 'steady state' science. Declining respect for such individual virtues as independence of mind, imagination, curiosity and intellectual fortitude is regarded as one of the most worrying aspects of the new regime.

For this reason, research councils and other patrons try to minimize the direct influence of competition for research funds on research careers. The policy of allocating resources to academic researchers strictly in terms of the scientific merits of their unsolicited proposals [§5.8] is designed to preserve many important features of academic freedom. Applicants for grants formulate their projects entirely in their own terms, and are judged accordingly by their peers. Indeed, the critical powers of peer reviewers are not inconsistent with the Mertonian norm of 'organized scepticism', and are usually exercised within a velvet glove of confidentiality, courtesy and responsibility. Every effort is made to keep the marketplace open for intellectual enterprise.

It must be said, however, that it is becoming harder and harder to square the image of the 'lone explorer of untracked territories of knowledge' (etc., etc.) with the grant applicant required to focus her research energies and goals into a series of uncertain yet over-determined project proposals, subjecting her hard-won skills, her life-long commitments and her fragile reputation to the superficial scrutiny and hurried decisions of an anonymous committee. Even if she is successful, there may be a suspicion – perish the thought – that she owes her reputation more to her skills at 'grantsmanship' than to sheer scientific capability. A scientist whose research career is continually being reshaped by the need to win further project grants cannot be considered 'academically free' in the traditional sense.

7.10 Managing

It would be an exaggeration to say that academic science is being systematically 'industrialized', or 'bureaucratized' [§1.3]. But universities and other research organizations are expected to be much more business-like in their internal arrangements. As we have seen [§5.10, §6.6] loosely organized academic departments are increasingly being redefined and restructured into self-contained institutes, centres, units, and other research entities. Like other 'small or medium-sized enterprises', these need to be administered, and directed. Then there are the research establishments and experimental facilities outside academia, not to mention all the funding bodies with their elaborate grant-awarding procedures to supervise [§5.8]. The trend towards more collectivized modes of research not only means that many researchers have to accept being *managed*: it also means that quite a few of them have to become *managers*.

This has a corresponding effect on the careers of scientists. The senior posts in universities are acquiring more and more administrative and managerial functions. The working day of a professor, for example, becomes overfull with a number of quite novel duties, such as making contacts with industrial firms and government agencies, obtaining research grants for his or her department, negotiating terms for the exploitation of research results, appointing staff on a series of short-term contracts, and ensuring that research projects are completed on time. Relationships with the higher echelons of the institution have also become a heavier burden, with much harder bargaining for a share of a smaller financial cake [§6.7], increased demands for compliance with safety regulations and fair employment practices, etc., not to mention formulating a departmental contribution to the corporate plan and compiling an impressive folio of staff research performance data.

Senior academics have always complained about all the administrative and committee work they have had to do. A Buggins' turn as department chairperson, Faculty Dean or Pro-Vice-Chancellor has long been the alibi for a temporary – or permanent – withdrawal from active research. But the convention was that such responsibilities were undertaken on a voluntary, short-term, part-time basis, and were to be considered additional and subsidiary to the more serious pursuit of a scholarly career. Although honorific in *CUDOS* terms, a senior academic post could be regarded as a public service to the academic community, in partial repayment for its support and recognition of individual achievement.

This aristocratic disdain for 'administration' is no longer fashionable. There is a definite trend in 'steady state' science towards the practice of industry, where the normal path of promotion is out of research as such and into the ranks of full-time management. Some of these duties can be performed perfectly well by professional administrators, but many of them require the sort of specialized knowledge and experience that can only be gained from half a lifetime of academic research and teaching. Thus, the traditional *path* of academic promotion from junior lecturer to senior professor has acquired much more significance as a career *ladder* up the ranks of a managerial hierarchy. Whether or not academic researchers are personally ambitious for advancement up this ladder, they know that this is the way to power and authority in the research system.

What scientists themselves say is that, the more creative a person, the more resistant they are to management and bureaucracy. This certainly applies to those who work best as self-winding individuals. But it also applies to many very great scientists, such as Ernest Rutherford or Howard Florey, who could direct and inspire the research of many junior colleagues. A more accurate statement might be that the administrative structures and procedures typical of large organizations are antipathetic to the extreme sensitivity and personal leadership that are required for the effective management of an activity such as research, which depends so much on motivating very strong-minded individuals to apply their capacity for curiosity and originality to problems that they might not have chosen for themselves.

This means that the people chosen to perform the more senior managerial roles in major academic institutions – Department Heads, Faculty Deans, Pro-Rectors, Deputy Principals, Vice-Chancellors, etc. – need to be selected and prepared for responsibilities which are not at all like those that arise in the performance or leadership of research as such. The experience of non-academic R&D organizations [§6.6] suggests a number of issues which have not yet been properly aired in academic institutions. What weight should be given to managerial capabilities in promotion at the middle levels of the system – for example, from reader, senior lecturer or associate professor to full professor? Conversely, what weight should be given to research performance in the selection of more senior academic managers? To what extent does academic research require a corps of professional managers and administrators with no serious research experience? Should all middle-rank research scientists be given formal training in management to prepare them for such responsibilities? Questions such as these are highly significant for scientific careers in 'steady state' conditions [§7.10].

The sharp distinction between *CUDOS* and *PLACE* careers has even been institutionalized. Many science-based firms have found that it is advantageous to give a high level of seniority to those of their abler scientists who still wish to stay active in research. They have therefore established a small number of prestigious, influential (and well-paid) posts for a few outstanding researchers outside the normal managerial hierarchy. A similar 'dual-career ladder' system operates in the scientific civil service, where there is often fierce competition for 'special merit promotion' to a senior rank without the corresponding responsibilities of line management. The same purpose is evidently served in universities by the increasingly common practice of promoting a very successful researcher on an *ad hominem* basis to a 'personal chair', outside the normal academic hierarchy.

Nevertheless, this duality is not yet fully integrated into the career aspirations of professional scientists. The role of the 'senior scientist' in an R&D organization is still evolving, while the managerial implications of achieving professorial status in a university are still underplayed. After all, these alternative ladders are designed for those few people who rise to levels where such choices become real. They do not provide motivation for the majority of scientific workers, who could perfectly well do some managing if they were given the chance, but who are also competent to remain active in research, if they so desire, until they retire.

7.11 Semi-employees and sub-contractors

From the point of view of an industrial manager, the traditional academic institution must seem an anarchic nightmare. There seems no way of actively mobilizing human resources to meet an urgent, radical challenge. Unlike industrial researchers following normal organizational careers, university staff in established posts are not customarily subject to directives concerning their research projects. Academic institutions tend to be organized on a collegial basis, where the administrative hierarchy has very little career leverage over its nominal employees, and is not able to deploy or redeploy them according to its own requirements.

This rigidity is not just due to academic tenure or to the obstacles to inter-institutional mobility. It is also inherent in the tradition of not accounting separately for the time that individuals spend on research and teaching [§6.9], and in the barriers to role changes, such as transfers between departments, within institutions. As every senior academic knows, personnel policy in academic research is principally a matter of appointing promising researchers in the first place, treating them generously when they have proven their worth, and providing them with tempting opportunities to change in desirable directions when these open up [§7.7].

Since the standing of a research institution depends primarily on the scientific achievements of the individuals who publish under its logo, a university has good reason to encourage its academic staff to do meritorious research and get it into print. This concern for the reputational careers of its employees ought to encourage institutional loyalty and coherence. It ought to give people the feeling that they can put themselves in the hands of their seniors and accommodate their research and career aspirations to the overall plans of the institution.

But people are ungrateful and perverse. From the point of view of the individual researcher, the university may seem little more than a managerial conglomerate, or a holding company, providing a secure financial base for the family firm [§5.10, §6.7]. One of the implications of improved accountability for research expenditure is greater transparency, and more cause for conflict over who has contributed, directly or indirectly, to what has been achieved. If a research entity does build up a reputation through which the university eventually gets a precisely calculable cash return, then the feeling is that this cash should be spent for the benefit of those who have laboured so hard to earn it. As with project grants and research contracts, the attribution of income-earning capacity to individuals or small groups can be a disruptive force within institutions that are struggling to hold together and act 'corporately' under 'steady state' conditions [§6.3].

This tendency towards fragmentation shows up clearly in the growing practice whereby an established academic researcher in a fashionable field of exploitable science builds up a profitable research enterprise on the side. There is the tempting possibility of pursuing an *entrepreneurial* career, where much the same scientific skills are used to earn a much higher (if less secure) income than the academic norm. This temptation is almost irresistible if these two careers can be followed in parallel, with the security of academic tenure to fall back on if private enterprise fails to pay off.

Small private-sector firms devoted to generating, exploiting or selling research results are a relatively new phenomenon in the natural sciences. Some observers attribute this phenomenon to the peculiar circumstances, first of micro-electronics and later of biotechnology, where progress in basic science was for a time running ahead of the R&D capacity of existing industrial firms. The major pharmaceutical companies, for example, were slow to recognize the potentialities of genetic engineering in developing new drugs, and thus opened up very risky but very attractive commercial opportunities for academic molecular biologists backed by venture capital. But this may have been a passing phase, since most of these start-ups have now closed down, or have been incorporated into the R&D operations of the big companies.

On the other hand, this development thrives in the general environment of 'steady state' science, with its ideology of maximizing the economic returns on research through market mechanisms. If, as many people argue, this is one of the best ways of ensuring that basic scientific discoveries are exploited technologically, then never mind any incidental damage to individual careers or institutional coherence.

Nevertheless, a parallel career system generates some awkward conflicts of interest – even of self-interest. Since universities are now in the business of selling some of their research capacity, their market strategies are confused by the activities of academic staff as commercial entrepreneurs in their own right. Thus, a contract by a university to provide a customer with a highly specialized research service may not be fully separable from a contract by a university employee to provide a similar service – even to the extent of using university infrastructural facilities – on his own account. Again, it can be a nice question for the ambitious researcher whether the latest hypothesis should be converted into a publishable research claim, which it is hoped will lead to academic promotion, or whether there is more to be gained by working it up quietly, outside academia, into a commercially profitable patent [§2.7, §7.6].

Academics and academic institutions are quite adept at combining and reconciling apparently contradictory roles, but this is a situation where satisfactory norms and rules have still to be established. In fact, the role of the semi-employee is not so novel as many people seem to think, since universities have long had comparable problems in regulating professional practice by academic lawyers, doctors, and engineers. There are excellent arguments for involving university teachers in the day-to-day work of the calling for which they are preparing their students, and, conversely, for drawing directly on the experience of research scientists, engineers, lawyers and doctors who are primarily employed in putting their knowledge to practical use. In other words, 'steady state' conditions in the academic marketplace strongly favour *hybrid* careers, which are the most effective means for opening up channels of influence between academic disciplines, between theory and practice, between science and technology, and between academia and industry [§9.7].

7.12 Formation

The human resources required for academic research take a long time to prepare. School work and undergraduate studies are only the initial phases in a very protracted process. What the French call the *formation* of an academic researcher stretches on into his or her late 20s or early 30s. There is no royal road to the technical knowledge and expertise required at the research frontier, even in a very narrow specialty. The question is whether the traditional mode of research training – especially the years of concentrated work and nervous effort that go into a doctoral thesis – is still the best preparation for a research career.

According to the *CUDOS* ethos [§7.4], there can be no substitute for the personal experience of conceiving, creating and presenting an original and critical work of scholarship. A doctoral dissertation is like the 'masterpiece' of a mediaeval artist: it is evidence of maturity in the art of research. Writing a thesis is not just a training exercise: it is the actual manufacture of a finished example of the type of product that the researcher will be putting on the research claims market in the life-long pursuit of a recognition. The work of preparing it is not merely a matter of getting the hang of some of the research techniques in a particular field: it is an all-round apprenticeship to a high calling.

No wonder it takes a significant fraction of a lifetime, and represents a significant fraction of a life's work.

Once upon a time, this lengthy and laborious process could be accepted as a normal part of a scholarly career. After a few impecunious years as a full-time graduate student, one could get a lowly post as a research fellow, instructor or junior lecturer while one was finishing off the experiments or field work, analysing the results, and writing it all up. In many European countries, this is still the normal practice. Scientists and other scholars are already professionally established in academia, often with tenured posts, before they complete all the formalities of a doctorate. These formalities may be very heavy, and may lock the candidate into a somewhat dated scientific problem for far too long, but they can be considered as ceremonial *rites de passage* marking progress along a predestined path.

Under 'steady state' conditions, however, basic training in the art of research is required over a much wider area of employment than academic science and scholarship. The science base is only one sector of an interconnected R&D system which uses the same technical skills and specialized knowledge for many different purposes, in many different contexts [§2.3]. And yet this system still relies on a single source – the research university – for the formation of its personnel. The experience of preparing and presenting a research thesis remains the unifying practice of this whole social institution.

We have seen enough of the extraordinary diversity of the roles now performed by research scientists to appreciate the difficulty of maintaining this tradition. Consider, for example, the convention that the subject of the dissertation should have been chosen by the candidate in the same spirit of academic freedom as any other research project [§7.8]. That is all very well in, say, a humanistic discipline such as history, where graduate students, like more mature scholars, work independently as individuals. But much research nowadays is teamwork [§3.7], where the student has to learn to act as a reliable cog in a social machine whose goals are beyond her control. Does a period of working as a very junior assistant in a high-energy physics 'collaboration' [§7.8] provide sufficient experience of science as a philosophy of nature, as a problem-solving method, as 'the honest search for truth', or as a self-governing republic of knowledge? Or is an elaborately contrived exercise requiring individual originality and scepticism an essential element even in the training of people whose careers will really be built on thoroughly professional technical expertise?

Some of the issues raised by the question 'What's the use of the PhD?' are of deep principle: others are very practical. For example, in many disciplines much of the actual work of academic research is done by doctoral candidates, guided more or less firmly by their supervisors. Only a small proportion of the students who pass through a graduate school eventually become established as university staff, but they provide a knowledgeable, dedicated labour force, easily exploited for building the reputations of their instructors [§7.7]. The whole structure of academic research would be upset if the locus of research training were to move out of the university – for example, into industrial R&D organizations, as some people suggest.

Although the debates on these issues often seem very parochial, they are symptomatic of the transition to a new regime. 'Steady state' conditions put great pressure on the leisurely, informal, although personally stressful practices traditionally associated with entry into the academic community. Many of these practices are of doubtful value in a world of more rapid scientific and technological change, closer linkages between discovery and application, more variety but less security of employment, more specialized division of labour but more collaborative, interdisciplinary projects.

The initial *formation* of research scientists is still one of the vital functions of academic science. But this is no longer dominated by the ideal of producing an individual piece of work that already satisfies all the necessary conditions for an original contribution to knowledge – that is, a book or paper that could be accepted for publication. The notion underlying this ideal was of a *moral* education, designed to develop traits of imagination, self-criticism, diligence, curiosity, devotion to truth, respect for the published literature, etc.

Scientists still need these personal qualities. But the balance of effort is moving towards a more systematic effort to train the aspiring researcher in the technical skills, areas of knowledge, and professional – even managerial – procedures likely to be required in the next stages of a productive career. This means taught courses and examinations, instruction in research methodology, experience with a variety of instruments and procedures, training in writing papers and preparing grant applications, and a general effort to broaden the vision of the student to include the applications and wider intellectual neighbourhood of his or her research topic. There are limits to what can be achieved by this shift from apprenticeship to deliberate training. Scientific research is a peculiarly subtle human activity which can only be learnt by doing it for real. Nevertheless, in this matter, as in so many other aspects of the subject, we must not be blinded by ideology and tradition to the realities of the world in which we are now living.

7.13 Mobility and versatility

In the past, the typical scientific career was a narrow path in a highly specialized research area within a self-contained academic discipline [§7.8]. A student of astronomy, for example, not only remained an astronomer for the rest of his days: he also worked more or less on his own until he became the world authority on some recondite topic such as the spectral properties of starlight, or the dynamical behaviour of galaxies. Increasing emphasis on multidisciplinary team research has not really reduced the premium on the peculiar knowledge and skills that take decades to acquire, but it is compelling such specialists to work very closely with their opposite numbers in other research fields or disciplines [§3.8]. The astronomer has to collaborate daily with the physicist who has invented a new type of spectroscope, or with a computer wizard who is re-analysing the data; she may even become quite at home in what was once an alien subject.

This deviation into interdisciplinary territories has to be serious; yet it may turn out to be short-lived. Scientific and technological fields are driven by internal and external forces into rapid growth or decay. Large research programmes are started up to meet urgent social needs, and institutions are founded to open up new fields of inquiry. The advance of the research frontier is now so rapid in some fields that research teams are not encouraged to become permanent administrative units assigned to particular problem areas. In tightly managed R&D organizations [§6.6], such teams are deliberately broken up and restructured from time to time, given new members with different skills, and directed towards new targets. As a result, individual career paths do not stay straight and narrow, but are forced to wander across the scientific map.

As we have seen [§6.7], such groupings and regroupings are not easily accommodated in the traditional departmental structures of universities and other academic research institutions. On the one hand, it is extremely wasteful of scientific talent to take on fully qualified researchers for each new project, and then throw them on to the job market like hired hands: on the other hand, tenured academic scientists strongly resist attempts to change their fields of research. New organizational entities are therefore being created within universities, with relatively large numbers of research workers covering a wide range of disciplines. Many so-called 'centres' are really 'circles' within which people may move intellectually from project to project while still retaining an umbilical connection with their disciplinary base.

In spite of the difficulties of maintaining scientific leadership and administrative coherence, programme funding and networking facilities may even make it possible to create regular communities for multidisciplinary cooperation between geographically dispersed institutions [§3.6]. It is quite common nowadays for the scientific career of a tenured academic at university A to pass frequently through a research laboratory in university B, and to be associated reputationally with a member of the staff of university C – often in another country [§8.10]. In other words, academic science is learning that the industrial practice of 'matrix management' can provide a continuing stimulus for individuals and small departments, and that directing a multidisciplinary project can contribute valuable experience to an academic career [§7.10].

Another dimension of career mobility is the way that university scientists are being encouraged – and rewarded materially – to become entrepreneurs or consultants in high-technology industry [§7.11], and scientists in government and research council laboratories are having to seek commissions for more socially relevant and urgent research. Again, if research resources are to be concentrated in centres of excellence [§6.8], many competent scientists will have to be encouraged to move geographically – or even be 'posted' to other institutions, as they could be if they were (like French scientists in the CNRS) all employees of a single organization.

Rapid cognitive and technical change, fierce competition for resources and for employment, frequent institutional restructuring, and other features of 'steady state' science may not only require scientists to be more labile in their research interests, and more mobile geographically. They may also have to be much more *versatile* in their skills. Instead of becoming and remaining narrow specialists in particular subjects or techniques, they may have to develop into 'problem solvers' or 'trouble shooters' over wide areas of science and technology. In other words, they are becoming much more like the 'QSEs' – the Qualified Scientists and Engineers – employed by industrial R&D organizations, where the condition for a *PLACE* is to be known simply as a reliable *expert*, rather than for trying to satisfy the disruptive academic norms of *originality and scepticism* [§7.4].

The science base is only one of a number of sectors of society employing a general body of QSEs. The boundaries are becoming less distinct between scientific research and other professions, such as industrial management, government service, technical administration, engineering, technological industry and commerce, teaching, advisory services, etc. People trained and qualified to be academic researchers are drawn or driven into other forms of employment at various stages in their careers. For this reason, concern about too few or too many people being trained for any particular research specialty are probably misplaced.

'Steady state' science is not a static, closed profession, providing a planned career for a certain number of carefully chosen, selectively educated and specially trained people: it is becoming a much more open and dynamic system, exchanging personnel with other systems as needs may dictate [§7.2]. The deplorable phenomenon of an excess of qualified researchers on short-term contracts [§7.11] could be interpreted as a sign of this latent dynamism. In the long run, the immobility induced by the 'freezing' of vacant permanent posts in academia is likely to be replaced by enhanced personal movement between competing institutions, between cognate disciplines, between academic and industrial science, and between more fundamental science and technology.

Career paths in 'steady state' science are thus likely to be more varied and less stereotyped than ever they were in the past. They will probably provide new challenges, new opportunities, and new roles for scientists and other scholars. It is likely to become more difficult to get into an established position very early with an outstanding individual contribution to knowledge – but easier at a later stage to exercise organizational influence and mobilize resources for an exciting collective attack on a difficult problem. There may be fewer openings for older scientists in a steepening hierarchy of formal authority, but more appreciation of their continuing value as active researchers.

Nevertheless, this fragmentation and diversification of scientific careers has its costs. It conflicts in principle with the personal commitment to a narrow field of study, and the long-term cultivation of highly specialized individual research skills, which have led to major scientific advances in the past, and which team research also requires. Some research projects – in forestry, for example, or in the harnessing of nuclear fusion – obviously have very long time-scales. In spite of accelerating change in our knowledge and capabilities, scientific and technological progress still depends greatly on the steady pursuit of very distant goals. This pursuit is difficult to maintain among scientists on short-term appointments passing through a series of temporary administrative structures with rapidly changing purposes.

7.14 Alternative career paths

Academic research under 'steady state' conditions is not usually short of 'human resources'. Looking up from the grass roots, the situation is typically described as a shortage of the material resources required to give research employment to all the available personnel [§6.9]. This situation looks quite different from the tree tops. The principal problem in funding research is then seen to be how to contrive that necessarily limited material resources do in fact go to the most capable researchers with the most promising research projects. In practice, this means ensuring that scarce resources are not wasted on researchers who, for one reason or another, are not likely to make good use of them.

The notion of individual selectivity is not envisaged at all in the *CUDOS* norms. Competition for esteem is the vital force of science, but the norm of *universalism* prescribes that all potential competitors should be free to prepare their contributions and submit them for critical assessment on equal terms. A procedure that cuts qualified entrants from the race, or severely handicaps them before it starts, is strenuously resisted by many academics.

From this point of view, the official rationale of the conventional system of project grants is acceptable, in that the principal criteria of choice are supposed to be the merits of the proposal as such [§5.4]. In practice, the scientific reputation of the proposer is usually a major consideration, but experienced referees know that they should not make this too obvious. Indeed, some funding organizations make the gesture of removing the names of grant applicants from project proposals in a vain effort to conceal them from referees – who then indulge in a guessing game to put the correct bias on to their reports.

What seems quite unfair is to deny a nominally qualified scientist – for example, an established staff member of a university – the opportunity to undertake the research that he is eager and competent to do, simply on the basis of his previous record. Nevertheless, this is the ultimate effect of any systematic procedure for the individual appraisal of academic staff in terms of their research performance [§5.5]. It is all very well to say that it is good for them to get an objective opinion on how they stand in the reputational market, together with paternal advice on how they might perform more effectively. The whole procedure would be valueless unless a substantial proportion of those who were appraised were told that it was very unlikely that they would ever eat lunch in that town again, and that they would be most strongly advised to spend their time elsewhere.

Of course this stark message would be conveyed within a velvet glove of sympathetic concern. Of course it would not have much operational force on someone with a tenured post. Of course it has always been the responsibility of senior academics, such as departmental heads, to watch the scientific progress of their juniors, and to reward success with promotion. But any formal procedure of this kind, which would be entirely normal in an organization run on *PLACE* principles, is just not consistent with the academic ethos. As we have seen, the notion of active personnel management, necessarily involving regular appraisals of individual performance and sanctions against those who do not come up to scratch, is generally regarded by the academic community as seriously out of keeping with its traditions [§9.4].

The reality of 'steady state' science is that it is uneconomic and inhuman to allow individual careers to drift in the rapidly shifting tides of scientific and institutional change. Universities and other science-based institutions are having to take more direct responsibility for the long-term welfare of their employees. They are also suffering external evaluation as institutions [§6.1], and are thus forced to take direct notice of the scientific achievements of all their staff, including those who actually do no research at all. However much this development offends against past practices, science-based institutions are introducing systems of career management, implying periodic monitoring of individual research performance followed by positive action to help people adapt to the changing circumstances and prospects thus revealed. Such procedures are undoubtedly advantageous for organizational reasons; they can also be very beneficial to individuals as a means of diagnosing and dealing with personal difficulties before they have become irreversible.

Career management is not required solely as a counter to the veiled *dirigisme* of bodies such as the UK Universities Funding Council, awarding research funds on the basis of research performance. It also has an important role in a more open market system of resource allocation. Strenuous competition is invigorating for those who are established and reasonably successful

[§5.9]. It is daunting for would-be entrants and demoralizing for the failures. A ruthlessly selective market system is not socially viable unless it has the means for limiting the wastage of human resources as people enter and leave the marketplace. This is where universities can play an important role as multi-functioned corporate bodies, responsible for the training, apprenticeship and early stages of professional employment of academic researchers, and for opening up fruitful careers in teaching, technical services, administration, etc. for those who have not been able to stand the heat of the research kitchen.

One of the abiding problems of scientific life is that people go 'stale' [§7.8]. Commitment to a narrow problem area can lead eventually to alienation from research itself. This is a situation where pro-active personnel management in the *PLACE* style can be extremely beneficial. It is accepted that corporate management should use its authority – and also provide opportunities and incentives – to move people out of dead ends before this happens, and to support them during the initial stages of working in a new field or in a different type of job. Indeed, experience suggests that highly specialized academic researchers may have a wider range of potential capabilities than they realize, and may adapt more quickly and effectively to challenging new problems than they would previously have thought possible.

One of the conventional pillars of active personnel management in R&D organizations is the argument that scientists do their best research when they are young. This may be true, but not because all scientists inevitably lose their aptitude for research as they grow older. Although the productivity of scientists does tend to decline with age, the general trend masks wide differences from person to person and from field to field. An older mathematician may find it more difficult to concentrate on a very precise problem, whereas an older biologist may have a larger stock of observations to draw on. Social and institutional factors, such as increasing managerial responsibilities [§7.10], also play a large part.

Even if individual productivity did decrease with age, the presence of at least a few older, more experienced, scientists in a unit can catalyse higher productivity from the younger members. A mature scientist apparently left behind by the advancing research frontier often has a valuable store of miscellaneous scientific information, and a tacit understanding on 'how to do research'. But if such people are to compete with younger, recently trained, recruits they have to learn their way quickly into quite new subjects. They may need a lot of help, including some formal retraining, to make this transition.

The point to emphasize, however, is that any process of individual selectivity is sure to generate serious disaffection among those who are effectively excluded from further research in mid-career, unless they can be offered attractive alternative employment. This is where the traditional multi-functioned academic institution comes into its own. A university is not just a research institute [§6.9]: its activities may include advanced scholarship, technical consultancy, professional practice, post-doctoral training, post-graduate teaching, undergraduate education, community services, etc., over every field of knowledge, not to mention its own administrative and technical services. Effective and sympathetic career management in academia must include the opening up of honourable career paths out of research into other activities of this kind.

Generally speaking, as a less rigid career structure is becoming customary throughout 'steady state' science, alternative career tracks are being opened up for those who leave research in their 30s and 40s. This runs counter to the tendency to make research more 'professional', and hence isolated from other callings, which can deter people from moving out of it gracefully, without loss of self-esteem. Note, for example, that the traditional terms of employment for scientists in many quasi-academic research organizations, such as government and research council laboratories, did not provide adequate opportunities for such changes, and were not consistent with the career patterns imposed by 'steady state' conditions.

Finally, let me say that this chapter does not pretend to cover every aspect of scientific careers. In particular, it excludes two issues of such importance that they really deserve whole books to themselves – the place of women in science, and the contributions of technical support staff to the research process. To tell the truth, I do not quite know what the effect of 'steady state' conditions has been on either of these serious but complex social issues.

For various reasons, science has never been an easy career for women. The hidden hand of cultural bias is slowly relaxing its grip, although it still remains strong. It is not clear whether the transition to 'steady state' science will facilitate this development. On the one hand, enhanced competition for employment is likely to make it more difficult for women scientists to return to highly specialized research after a break. On the other hand, a general career structure where there are more frequent changes of job and more diverse professional roles may give them more scope for advancement than at present.

A noticeable feature of scientific institutions is the very high stability of employment of non-graduate technical staff. The longest-serving employees of many research establishments are usually not the scientists, but the skilled technicians. The increasing sophistication of research apparatus that is characteristic of 'steady state' science may reduce the ratio of support staff to research scientists, but may require them to be educated to much higher levels. In spite of the strong demand for their services, they may need much more training and retraining to maintain high standards of expertise in a rapidly changing technical environment.

8
Science without frontiers

I had a little nut tree, but nothing would it bear,
Except a silver nutmeg and a golden pear.
The King of Spain's daughter came to visit me,
All for the sake of my little nut tree.

8.1 The multinational tradition of science

On ceremonial occasions, scientific notables congratulate each other, and themselves, on being members of a world-wide community that has always 'known no frontiers'. This is an admirable sentiment, but what does it mean in practice? Paradoxically, on other occasions, the same scientists say that science nowadays is 'going international'. This apparent inconsistency is yet another manifestation of the transition to a new regime in science.

It is quite true that science has always been a *multinational* activity. It has never been realistic to consider science in the UK independently of science in other countries, particularly its European neighbours. One of the features of the Renaissance was the rapid diffusion of new scientific discoveries and theories across the Continent, carried by scholars, churchmen, artists, craftsmen, mountebanks, diplomats and other travellers, as well as by printed books.

Since the seventeenth century, scientists have sought international recognition for their discoveries. The European market for scientific ideas set both the goals and the standards for research. Scientific journals, such as the *Philosophical Transactions* of the Royal Society in London and the *Comptes Rendues* of the Academie des Sciences in Paris, were published nationally, but they circulated freely and were cited indiscriminately throughout the scientific world. Isaac Newton, for example, felt the necessity of publicly defending himself from the French and Italian opponents of his astronomical and optical theories, as well as from the claims of the German, Leibniz, to have forestalled his mathematical inventions.

The scientific elite of the leading scientific nations were often personally acquainted, both as colleagues and as competitors, and were linked informally through their learned societies. There is a famous story of Sir Humphrey Davy, accompanied by his brilliant assistant Michael Faraday, on an honoured scientific visit to Paris at a time when England and France were officially at war. On the other hand, scientific competition between individuals has often been associated with scientific rivalry between nations, especially when these were also competing economically and militarily. Leading scientists of every country, in every age, have been involved in the design of novel weapons of war. In spite of an enduring tradition of international conferences, personal travel and foreign employment, controversy over the standing of 'British science' relative to 'French science' or 'German science' dates back several centuries, and has often erupted into bitter quarrels over priorities and prizes.

As a matter of fact, the historical record on 'science without frontiers' is far from perfect. It is often forgotten, for example, that German scientists were shut out of the international conference circuits for years after the First World War. Nevertheless, even in a ferociously nationalistic world, academic science has always tried to practise seriously the universalism that it preaches. At the height of the Cold War, for example, scientists on both sides took professional notice of each other's publications, even if they were not permitted to cross the Iron Curtain to discuss them together. One of the first channels of communication to open up across this great political divide was the meeting of leading scientists from both sides at Pugwash, Nova Scotia, in 1957.

This universalism is a fundamental constituent of the scientific ethos [§7.4]. There are unassailable reasons why science needs to keep intellectually open across the world in this way. But, as we have seen in previous chapters, the scientific ethos is also intensely individualistic. This individualism is not glorified as such, but is very obvious in the public recognition and celebration of personal achievement at all levels, from the PhD degree to the Nobel Prize. The wider the scope of this recognition, the greater the achievement. For this reason, scientists of all countries are strongly motivated to compete as individuals in the global knowledge and reputation markets of their various disciplines [§7.3, §7.5].

Notice, however, that the coveted status of being an 'acknowledged international authority' on a particular subject is informal and personal. It means having one's books translated into foreign languages, or being invited to give the opening address at an international scientific congress, or being elected to a foreign scientific academy. It is not necessarily connected with any transnational organisational function, such as chairing an agency of the United Nations, negotiating on behalf of one's country with the European Commission, or sitting on the board of a multinational company.

Being a member of the international scientific community is not a very distinctive social role, and not in any way inconsistent with the rights and responsibilities of a citizen of a particular nation state. Scientists have often been accused by nationalist extremists of being 'cosmopolitan'. They have always welcomed the opportunities for meaningful travel, and the linkages of colleagueship and friendship with people of similar specialized interests in other countries. As the author of paper, as an expert referee, or as a participant in an international conference, the academic scientist dons a universalist garb, which he or she may well find much more comfortable and more becoming than that of the cocksure patriot. But the formal scope of this cosmopolitanism is limited to the specific activities of the 'scientific world'. When the guns begin to shoot, as the tragic history of the twentieth century amply confirms, scientists are as ready as the rest of humanity to follow their national flags into war.

8.2 Voluntary associations of invisible entities

Scientists have always insisted that they belong to a world-wide community engaged on a common enterprise. By the 1930s, systematic transnational scientific activity was taking place on quite a substantial scale, primarily on a voluntary basis. Scientists exchanged information, travelled widely and met together, either at their own expense or through the patronage of charitable foundations. As shown by the history of the growth of new research fields, such as atomic physics or biochemistry, even quite junior scientists in Europe and America became acquainted with each other as fellow members of the 'invisible college' of their discipline or specialty. But this was typically an indeterminate social group which continually eluded institutional solidification.

In spite of its communal ethos [§7.4], science was very weakly organized transnationally. Until after the Second World War, there was very limited direct collaboration on research projects, or programmes, except where this was essential for practical purposes such as weather forecasting, telegraphy or the standardization of weights and measures. The occasional multinational research project was 'coordinated' by independent national scientific bodies acting in concert, rather than 'managed' collectively from a transnational centre. The 'foreign relations' of established scientific institutions, such as universities and learned societies, were often warm and harmonious, but they were very seldom made precise in legal terms or cemented financially. Multinational scientific organizations such as publishing houses, research institutes and academies scarcely existed 'internationally' in a formal sense. They could only be incorporated legally under national laws, and were sometimes very weak even at the national level.

For academic scientists, the universalism of science is an ethical norm. But the idealized notion of a world-wide 'republic of science' is always linked to the 'purity' of its operations [§2.6]. Knowledge that had been acquired 'for its own sake' could be deemed 'useless', and hence could safely be made public. Indeed, the human race as a whole would benefit by making it available to anyone who might later think of how to profit from it [§2.7]. But as this knowledge moves out of the academy into the commercial world, this norm ceases to apply. A century ago – in the era of Nobel and Edison – a new institutional form of research was developing. As we have seen [§7.4], the science applied in the industrial or governmental laboratory was *proprietary* and *local*. Research results were the intellectual property of the scientist's employer [§7.6]. They might or might not be kept secret, but they were not originally sought with world-wide publication in view.

Until quite recently, most research organizations run on *PLACE* principles were geographically localized, with almost no transnational connections. Even in industrial firms involved in international trade, R&D was normally carried out in the headquarters laboratories of the parent company, in its own country. Scientific work was thus identified with an organizational role which had no significance in a larger world.

There were, of course, a few very large firms whose scientific experts might have to visit foreign subsidiaries. There were large colonial empires where a military or civilian scientific official might be employed on a roving commission. But the work itself, in the laboratory or design office, was not embedded in a supposedly universal body of knowledge: indeed, an overtly cosmopolitan attitude might give rise to suspicions of disloyalty to the firm or the nation. Parochialism was the prevailing attitude.

8.3 The globalization of R&D

This peaceful scene has changed dramatically in the past fifty or so years. Science is 'going international' in a big way. Immense experimental instruments at Geneva, Hamburg and Grenoble – even at Culham, in rural Oxfordshire – are served or used by researchers and technical staff from all over Europe [§3.6]. Space agencies fly scientific satellites designed by multinational teams. Astronomers make direct observations by telecommunication linkages to telescopes on the other side of the world. Earth scientists and ecologists from a dozen nations take part in research projects covering whole continents or whole oceans. International institutes for molecular biology, systems analysis, and plant breeding are located in, and employ staff from, the First, Second, and Third Worlds. The budget for the R&D programmes supported directly by the European Community runs into billions of ecus. Multinational firms spread their research efforts over a web of centres in Europe, North America, and Japan. Heads of national research councils consort together to draw up treaties and contracts for collaborative research. And the scientists themselves seem to spend more time in foreign hotels and aeroplanes (economy class, though) than in their laboratories, libraries or class rooms.

These are the quantitative signs of a qualitative change in the nature of the scientific enterprise. In effect, the traditional *cosmopolitan individualism* of science is rapidly being transformed into what might be described as *transnational collectivism*. The ethos of the old regime still operates on the surface, but new institutional structures, created by typical 'steady state' pressures, are now beginning to emerge on this larger scene. Most of these, are 'multinational', in the sense that they have normal legal and managerial roots in one or more of the countries they cover: but some, like the agencies of the United Nations, are truly international, in that they exist in the floating world of organizations that claim to be free of allegiances to any specific country.

The forces driving this transformation are easily identified. Some of these forces are economic and political. World trade in science-related commodities has intensified and become more competitive commercially. As a result, the world markets in scientific knowledge and technological know-how have become more comprehensive and more closely connected [§7.5]. A pharmaceutical researcher in Switzerland must know immediately all about a new drug discovered in Japan: a Japanese electronics engineer keeps a close watch on the results coming out of a physics laboratory in Massachusetts: the claim by scientists in Utah to have observed 'cold fusion' has to be assessed very seriously in London, Paris, Frankfurt and Tokyo, as well as in New York and Washington.

As a result, many multinational companies have been induced to extend their R&D activities from their parent countries and to organize them internationally. In some cases this may involve no more than setting up a laboratory in a foreign country to monitor the technical operations of a local manufacturing or marketing subsidiary. In other cases, a multinational R&D organization has to be set up to run a multinational industrial programme, such as the design of an aircraft or weapon system.

The largest multinational companies in science-based industries such as pharmaceuticals have gone further. For a variety of reasons, such as tapping local sources of highly skilled scientific personnel at relatively low wages, or gaining access to a distinctive academic or technological culture, they find it advantageous to have major corporate laboratories in several countries. Some of the most important industrial R&D laboratories in the UK belong to American or European multinational corporations such as IBM or Hoechst, while several large British firms have major research facilities in Europe and the United States. In the last few years, this globalization of industrial R&D has spread to the third major region of the developed world, centred on Japan, and there will surely be similar extensions into the former Soviet Bloc region and into Latin America. Scientists whose working arrangements were previously quite parochial are thrust into a transnational organizational environment, which puts them into regular contact with their opposite numbers in other countries, and may require them to live abroad, or move from centre to centre within the same company.

Strangely enough, commercial rivalry between industrial firms does not prevent them from cooperating internationally in some forms of R&D. Sometimes this cooperation takes place under the umbrella of an intergovernmental body such as the European Community. Thus, for example, ESPRIT (the European Scientific Programme for Research in Information Technology) extends to the whole of Europe the spirit of the UK Alvey Programme, which involved a number of competing industrial firms. But it is not unusual nowadays for two or more multinational firms to collaborate on the R&D required to bring a technological innovation, such as an ethical drug, nearer to the market.

The working arrangements for such programmes are as complex and diverse as the commercial, political and technical conditions in which they have evolved. In some cases, for example, reference is made to technologies that are in a 'pre-competitive' stage of development, indicating that proprietary rights would not be claimed for the immediate research results [§2.5]. This is the formula under which a great deal of collective R&D effort by European firms is fostered by the EC, ostensibly to put them in a position to compete more effectively with the Americans or the Japanese. In general, however, such arrangements are probably best understood as parts of much broader schemes of industrial, commercial or military cooperation. The increasing globalization of industrial R&D is less an innate tendency than a corollary of the globalization of markets in advanced technological products.

8.4 Multinational science for multinational communities

In spite of every effort of Thatcherite and Reaganite politicians to make it seem so, not all 'applied' science is commercially competitive. Proprietary rights are not the primary consideration in R&D directed towards the improvement of public, non-commercial goods, such as public health, environmental safety, welfare facilities, education, transport and energy supply networks, and so on. As states agglomerate into international communities, communal research fans out multinationally. The Chernobyl disaster demonstrated dramatically that 'our common European home' is just a single region for the purposes of much environmental research. There is increasing realization that certain political initiatives, such as the struggle against increasing pollution of the oceans and atmosphere, require a global approach, backed by world-wide scientific research. Again, certain health problems are common to so many countries, especially in the Third World, that it makes no sense to study them in separate national programmes.

Although much communal research – for example on climatology – is really more strategic than applied, it is seldom bounded by national frontiers. This is becoming more and more evident even in the social sciences, where cultural diversity plays such an important part. Thus, for example, all the practical arguments for supporting a national survey of, say, public attitudes to science and technology extend automatically to comparative studies of the same attitudes in a number of countries, preferably according to the same research protocols.

As we have seen [§5.7], the individualistic practices of academic science cannot be relied on to ensure that the communal research requirements of any one country will be systematically and reliably met. That is why we have the national research systems that are so characteristic of 'steady state' science. It is now quite obvious that the need for a collective approach applies with even greater force to the communal research requirements of world regions – indeed, of the whole earth. Even if every country were very energetic in trying to fulfil its own scientific needs, some very important (if not necessarily enormously expensive or large-scale) research might only be undertaken very badly, if at all. In other words, scientific programmes that transcend national frontiers have to be organized, at least to some extent, by bodies that transcend purely local interests.

This trend towards transnational programmes of communal research, whether directly applicable or more indirectly strategic, shows clearly in the R&D portfolio of the European Community. This portfolio not only contains environmental programmes where Europe can be considered a single geographical entity. It also covers strategic projects on themes such as energy supply, whose objectives cannot be achieved by one country 'going it alone', either because they are beyond its human or financial capacity or are likely to be worth while only on a Europe-wide scale. And, as the world approaches the twenty-first century, more and more economic and political muscle is being given to bodies charged with developing global research programmes, on global issues such as climate change, environmental degradation, and ecological conservation, to serve a global community.

Of course, there are only vague boundaries between communal and industrial R&D, and between strategic research and the marketplace. This becomes evident as soon as a high-minded programme of international scientific cooperation seems to be opening up opportunities that might be exploited commercially. As the controversy over patenting the DNA sequences discovered in the Human Genome Project shows, this basic research programme then has to be treated as if it were a preliminary phase of a technological development whose results are likely to be appropriated by the nation or firm doing the research [§2.7]. The possibilities for international coordination and selectivity then depend on the wider issues of international industrial competition and cooperation noted above. Unfortunately, the point at which it would be advisable to make this transition to a different mode of coordination is seldom well defined.

International scientific cooperation is not entirely driven by practical necessity. It is often motivated by quite extraneous political considerations, such as the desire to demonstrate international goodwill or as one element in a general agreement for economic, industrial or military collaboration. The high cultural profile and the distance from commercial considerations of academic 'Big Science' [§3.6] makes it very attractive to governments as a symbolic medium of international cooperation.

Governments sometimes agree to scientific exchanges and joint research projects in much the same spirit as they arrange exchange tours of their ballet companies – that is, essentially for sacramental reasons. Indeed, the peaceful image of science is often exploited symbolically. The announcement of such an agreement in the wake of a fruitless Cold War summit meeting was startlingly reminiscent of the customary exchanges of armour, horses and slaves between mediaeval potentates on similar occasions of scarcely veiled hostility. Not surprisingly, the scientific quality of the multinational research projects engendered in this way is usually so low that they would not qualify for funding on normal scientific grounds, and are only supported for diplomatic reasons.

8.5 The growth of a transnational science base

As we have seen in earlier chapters, 'steady state' conditions have stimulated two major developments at national level. On the one hand, every government has had to set up a policy apparatus to determine priorities, allocate limited funds to competing research institutions, and initiate research programmes for public benefit [§5.1]. On the other hand, political, financial and managerial resources have had to be mobilized to create major research 'facilities' on a much larger scale than could have been undertaken by any existing academic institution on its own [§3.6].

The twin-pronged rationale of collectivization does not stop at the national level. On the one hand, governments support multinational research programmes to tackle transnational problems. On the other hand, it has become increasingly obvious that the specialized division of labour in *academic* research can be as advantageous internationally as it is supposed to be nationally [§3.9, §6.8]. In particular, the economies of scale in the provision of expensive, indivisible, research instruments are even greater if the costs can be shared by many countries. Certain fields of basic research – notably high-energy particle physics – have generated such intense pressures for the aggregation and concentration of resources that they have burst out of their national frames and gone totally international.

The image, and also the prototype, of transnational 'Big Science' is, of course, CERN [§3.6, §7.9]. In principle, this is a tenuous organization for spending collectively a part of the science budgets of a number of European nations. In material terms, it is identified with the enormous particle accelerators, the workshops, the computers, the lecture rooms, the technical and administrative staffs of a single enormous institution sited in Switzerland near Geneva. To describe it as a 'facility' is to undervalue the variety and complexity of its scientific resources. Thus, it houses a number of instruments, built successively over a period of 40 years, each prodigiously exceeding its predecessor in size, scientific power and cost. It is also the nerve centre of an international web of scientific activity, ranging from the design and construction of the components of immense experimental rigs, to the theoretical analysis of research data. In effect, it is the home institution for a large scientific and technical community engaged in the study of a major aspect of the natural world.

CERN is outstanding as an example of the internationalization of science. It is remarkable in its scale, in the fundamental nature of its research goals, in its relative independence of political or commercial forces, and in its scientific achievements. But it is only one of a number of major research 'facilities', supported and managed internationally to serve basic science around the world. At some of these facilities the actual experiments are carried out, as at CERN, by very large research teams that are themselves international. The members of such a 'collaboration' [§7.9] are supposedly independent scientists, academically based in many separate universities in a number of countries. In practice, the research has to be carefully planned and firmly directed, with all that this implies for work organization and quality control.

High-energy particle physics is unique in the indivisibility of both its big machines and its big experiments. In space research and astrophysics, the central instrument, such as a ground-based radio telescope or an orbiting satellite observatory, may be very elaborate and very expensive to operate, but it is often capable of performing a number of more or less independent scientific observations or experiments. The research programme can then be broken down into a number of different projects, to be undertaken by relatively small research teams [§3.6]. This still means, however, that the main instrument has to be funded, built and operated on a transnational basis, and that the various scientific projects for which it is used have to be designed and programmed to fit snugly together in time and space.

This very close coordination of instruments and research protocols is not necessary for basic scientists using some major international facilities, such as the neutron and synchrotron radiation sources at Grenoble and Daresbury. In effect, these are rather like very large computers or databases, with many independent work stations where small research groups can run their experiments simultaneously. A multinational organization is still needed to construct and operate this instrument, but access to it can be decided in the normal way by national funding bodies, each with a fixed quota for its own clients. In other words, for most of its users, such a facility is a powerful means for doing 'little science', although it may involve them in much travel and day-to-day contact with the foreign scientists also attracted to the scientific honey pot.

Indeed, so much of the talk about the organization of science internationally is focussed on 'Big Science' that we tend to forget that most academic research is not carried out by large teams of scientists and engineers using gigantic instruments to perform a few dramatic experiments, each costing millions of pounds. Nor is it like technological development, where hundreds of technical personnel have to collaborate closely to cover every aspects of the design of a new commercial product such as a computer chip or a drug. In the academic laboratory, 'little science' still rules. Much basic and strategic research can still be done effectively, efficiently, and to a high standard of excellence, in relatively small research entities [§5.10].

Experience has shown, moreover, that satisfactory arrangements can be made in an advanced industrial country to site such entities in suitably diverse intellectual environments, such as major research universities, and to provide them with most of the other material and human resources they need from local or national sources [§6.7]. Whatever the political arguments for aggregating researchers into large, specialized, international institutes, these are not technically necessary for ordinary laboratory science, provided that arrangements can be made for occasional access to a very big instrument such as a telescope or specialized radiation source, or for participation in a major international research programme such as the Human Genome Project.

The principal driving force for international collaboration in most fields of academic research – and this includes the social sciences and humanities – is still the norm of universalism. The traditional national markets in knowledge creation and communication have increasingly been globalized [§8.3]. This is clearly shown by such indicators as the growing number of scientific papers written jointly by authors from different countries and the increasing proportion of citations, even in patent applications, to 'foreign' authors. Scientific databases and publication archives are run as world resources. Many scientific journals are actually published internationally, both by learned societies or commercial publishers, and are governed by multinational editorial boards.

To a very large extent this is because the technical progress of science in every field has become dependent on the rapid transfer of data around the world. Multinational research has been both facilitated and fostered by improved and expanded means of communication and travel [§3.6]. More convenient international telephone and data-transfer links and cheaper air travel have made it possible for scientists in different countries to work together on specific research projects, almost as if they belonged to the same university. Thus, the scientists of the former Soviet Bloc have fallen far behind the research frontier in many fields of 'little science' because of the lack of direct access to the high-powered computer networks and searchable databases that now shape the whole research environment in the West.

The links established for the rapid transfer of technical data are more and more being used for interpersonal communication. In high energy particle physics, for example, the only really significant scientific meetings have long been the periodic international congresses where all the active researchers expect to meet and exchange informal information about their work. These meetings continue. But the electronic data links spawn electronic mail networks enabling scientists from several countries to work together on a day-to-day basis on a single experiment. This sort of collaboration, which is also now quite normal in many 'little science' specialties, is often organized quite informally, without reference to any official arrangements for international scientific cooperation.

8.6 Programming research internationally

Academic science nowadays is thus becoming ever more cosmopolitan. Researchers require almost instant access to databases and publication outlets in order to keep up with their subjects. They link themselves around the world through E-mail into real-time communication networks, and jet back and forth across the globe to establish frequent face-to-face contact with their specialty peers. They take jobs in other countries, for shorter or longer periods, to gain wider experience and expertise. Even if they are working on 'little science' projects, they fly off for a week or two to a major research facility, such as a synchrotron radiation source, where they share a work station, and eat in the same canteens, with researchers from all over the world.

In a world where many of the political barriers to personal mobility and interpersonal communication have been torn down, it is easy to understand how it is that scientists are becoming more cosmopolitan as *individuals*. But the organizational units of modern science are not individuals, but *groups* [§3.9]. In many ways, an academic research entity does behave like an individual. For example, it is in the market for reputation [§7.3], and cannot be considered effective unless it is contributing to, and is recognized by, the 'Invisible College' of its specialty. But even quite a small research entity, with a dozen or so members, is not equivalent to the unfettered individual envisaged in the academic ethos [§7.4]. As we have seen, it is a peculiar type of social institution, embedded in a matrix of local academic, political, cultural and commercial obligations [§5.10]. New administrative and legal arrangements may be required before it can become 'cosmopolitan' in its working practices.

The scientists who lead research entities are in two minds about how these arrangements should operate. On the one hand, they are united in their support for cosmopolitanism as such, and appreciate fully the case for collective, multinational research. On the other hand, they fiercely defend the scientific autonomy of their research groups – which they interpret as their own personal academic freedom [§7.9] – and are strongly against the systematic 'internationalization' of their work. They accept that the science base has become increasingly interconnected globally, and that international scientific *cooperation, coordination* and *collaboration* are highly desirable, but they would insist that research in most fields does not need to be put under international *management* [§5.8]. The pressure for this comes mostly from elsewhere in the research system – typically from politicians and civil servants who argue for the 'rationalization' of science policy at a supranational level.

What the scientists favour is the old-fashioned 'bottom up' process whereby independent researchers (or, nowadays, research entities) agree voluntarily to work together. As we have noted above [§8.5], this is already happening transnationally on a large scale, without any need for official encouragement. National research funding bodies have sometimes found that their rules did not envisage support for projects undertaken jointly by research entities in several countries, but they have been quick to appreciate the virtues of such collaboration, and to make arrangements with foreign funding bodies to support such work jointly. Indeed, many of the R&D programmes of the European Commission only accept joint applications from research groups in two or more countries – especially countries with underdeveloped scientific capabilities such as Greece and Portugal – in order to foster regional 'cohesion' in science.

The logic of the situation for most academic scientists does not drive them to come together spontaneously to coordinate their research internationally and to develop new international research programmes. But the international scientific marketplace is crowded with specialist conferences where scientific leaders report progress in their research and inevitably discuss over wine and coffee the possibilities of closer cooperation. A significant role in furthering these discussions is still being played by the traditional international institutions of academic science – the learned societies, headed by the International Council of Scientific Unions (ICSU). The European Science Foundation (ESF), which is a non-governmental club of the quasi-governmental bodies funding the national science base in each European country, brings together the grandees of academic science for just such purposes. From these low-cost, loose-linked, idea-swapping bodies have come the initial plans for such important developments as the International Geophysical Year, the International Biosphere Programme, coordinated Antarctic research, and the European Geotraverse.

Here, as always in science, there is a tension between cooperation and competition [§9.9]. The scientific community strenuously resists any development that might inhibit the public competition between rival research centres, especially when this rivalry is fuelled by considerations of national prestige. Yet experience in recent years has shown that it is not too difficult to persuade academic scientists to coordinate their individual research efforts if this is seen to be the only means of their obtaining the collective benefit of a powerful new facility or access to a wider field of scientific observation.

CERN is the prime example of a research facility which came into being through the active efforts of the scientific community. Back in 1950, a small group of eminent physicists from the leading nations of Western Europe already saw clearly that the experimental apparatus that they needed for the next leap forward in high-energy physics was beyond the means of any one of their countries [§3.6]. Despite the economic difficulties of the immediate post-War period, their governments were coaxed into clubbing together to provide funds on a scale to challenge growing American predominance in this field, and eventually to re-establish Europe as an equal partner in this fundamental branch of science.

The CERN example has since been followed in many other fields of science. In astronomy, for example, researchers from all countries accept that international collaboration to share the costs of expensive research facilities such as the European Southern Observatory has become the norm, and they lobby hard to get the funds for these facilities from national and international sources. That does not mean that they favour international institutes directly employing large numbers of expatriate researchers. Their preference is still for user groups to remain firmly based in their home universities and other research institutions [§8.5], although they accept that these may have to be organized and run as interacting components of an international research machine. In effect, the members of various specialized subsections of the scientific community willingly relinquish control of their local or national research efforts in return for access to much more powerful research capabilities.

Continued reference to the European region as a model for such developments is inevitable. The scientific preponderance of the United States in the second half of the twentieth century seriously unbalances the global market in academic research. The sheer size of the Soviet Union made it seem a potential competitor, but it never really matched up to the US in basic science, and its over-blown national science base has now collapsed. As we have noted in the case of CERN, much of the pressure for international collaboration in science arises from the desire of European scientists for their own research facilities comparable to the 'Big Machines' funded nationally by the US Federal Government. This desire to establish 'European science' as an equal partner (or worthy rival) to 'American science' is a force in its own right. Indeed, the establishment of a European Molecular Biology Laboratory (EMBL) in Heidelberg in 1977 owed more to this sentiment than to the technical demands of the subject.

At this point, however, the discourse has moved from the domain of science to the domain of national and international politics. This is unavoidable. Any field of research that is 'going international' is bound to have been thoroughly collectivized already at the national level. Its scientific spokespersons will not only be senior professors and academicians: they will also be chairpersons of research councils, members of government advisory bodies, directors of large institutes, or even senior civil servants. Although they may still be active in research, and have their feet still among the grass roots of their community, they are familiar with, and operate in, the tree tops of state power.

Not surprisingly, therefore, the field of action has moved towards the intergovernmental organizations, such as the European Community, which wield this power transnationally. The 'top down' forces that shape 'steady state' science nationally [§5.2] are now being exercised from a higher level still. Most of these forces are facilitative rather than coercive. Academic research groups working on similar problems in, say, the UK, the Netherlands, and Portugal do not have to collaborate, but the funds they might win for a joint project from the Framework Programme of the EC are a juicy carrot. Controversial decisions about the size and shape of this Programme in the Council of Ministers may not yet have much effect on national research budgets or priorities, but they do mean that many issues concerning communal research are aired in regional terms. Governments that have become accustomed to spending quite heavily on research in agriculture, environmental protection, transport, energy supply, medical services, etc. do not have to reconsider their support for such activity, but they do have to take into account the large transnational programmes being organized and funded from Brussels in these fields.

The Community, and similar bodies such as the European Space Agency (ESA), are under other pressures to become proactive in the field of R&D. The complex, often unacknowledged, linkages between industrial firms and national governments extend upwards into multinational space, and encourage international agencies to support very large strategic research programmes [§2.5, §5.8]. Where the science relates to a technology that is itself fully internationalized, such as satellite communications or fusion power, the corresponding international agency essentially calls the shots, and may establish its own research establishment – for example, the Joint European Torus at Culham, in Oxfordshire – to pursue the subject intensively.

The stakes in the international science game are increasing. Space research has long been too expensive for Europe to undertake on its own, without access to US or Soviet satellites. High energy particle physics is also beginning to transcend the funding capacity of any one continental region, European or North American. Some research sites, such as the oceans and the atmosphere, are so extensive that their study requires collaboration on a global scale. This means the development of international organizations to foster, support, coordinate, or manage research on that scale. It now seems inevitable, for example, that research in high energy physics will be organized on a world-wide basis, around a single 'very big machine' [§3.6]. The scientific community in this field is already integrated globally: the transnational political machinery will soon have to be set up to run this facility.

8.7 The organizational kaleidoscope

What we are seeing, then, is a strong trend towards the *transnational collectivization* of science. The economic and technical considerations favouring such a trend in particular cases are often extremely cogent – sometimes absolutely compelling. There is no way, for example, that the individual European countries now sharing the costs of CERN could have remained in high energy physics research except by moving to the transnational level. And there is no way that the work of such a large and complex scientific facility could have been organized without setting it up as a relatively self-contained formal institution with an integrated managerial structure.

A growing proportion of world scientific activity is now organized multinationally. It is becoming less and less realistic to talk about the 'national science base' of a country such as the UK as if it were a distinct administrative or political entity, albeit operating in an international environment. As we have seen [§8.5], a great many British scientists are doing most of their research at international research facilities, such as CERN, or they are employed by international research institutions, such as EMBL, or they are being funded our of international research programmes, such as ESPRIT, or they are collaborating in international research projects, such as the European Geotraverse. In addition, a great many more are employed by multinational companies. When the direct or indirect cost of these activities are totted up – as they are by zealous Treasury officials – they amount to a sizeable fraction of national expenditure on science.

The European Community inevitably figures largely in these developments. Whether or not it eventually evolves into a 'United States of Europe', it is beginning to have a significant influence on UK science policy. So far, the overall R&D budget of the EC is only a small percentage of the total national R&D expenditures of its member states, but it has been rising so rapidly in some fields that it is now a major source of funding for academic research in the UK. Indeed, senior academics are becoming quite worried that their institutions may be beggared by the *indirect* costs [§3.5] they have to bear when the leaders of their research entities triumphantly report that they have won a magnificent project grant from Brussels.

The European dimension of UK science policy is actually much more extensive than 'Europe' in this narrow political sense. 'Multinational science' is not, of course, a homogeneous activity. It varies greatly in scope and coherence. Since the early 1970s, a bewildering variety of organized bodies have emerged to put it into practice. The institutional network that has thus evolved is inevitably more complex and disconnected than in any single national system.

Even at the top end of the scale, the administrative arrangements governing 'Big Science' facilities are quite diverse. Some of these, like CERN, are intergovernmental consortia with very large budgets to which member governments subscribe by treaty. They employ their own staff, and have established a tradition of being largely under the control of the scientists who use their facilities. Others, such as the Joint European Torus (JET), are mere subsidiaries of much larger, politically ponderous, intergovernmental bodies such as the European Community. Others again – for example, several UK astronomical instruments – are national centres, whose facilities are rented out on a regular basis to research teams from other countries.

The administrative arrangements for multinational collaborative research programmes, such as ESPRIT [§8.3], are much more skeletal. Quite often these consist of little more than collections of projects funded under a common heading from an international budget but actually undertaken quite independently by various research groups in different countries. In other cases, however, each project has to further the cause of 'cohesion' by involving the direct collaboration of researchers from several countries, although not necessarily in the same laboratory.

Whether or not this collaboration is scientifically efficient, it does foster intergroup cosmopolitanism by generating transnational networks of research groups, who keep in regular communication and are encouraged to collaborate informally as opportunity offers.

Many of these international developments originated in Non-Governmental Organizations (NGOs) such as the ESF and ICSU, who use their limited budgets to coordinate scientific work internationally [§8.6]. In many cases NGOs have been the channels through which national governments have chosen to act internationally, although in some fields this role is now being taken over by more tightly organized intergovernmental bodies such as CERN or the European Space Agency (ESA) – not to mention one or another of the Directorates-General of the European Community.

The United Nations 'family' of organizations includes several, such as UNESCO, WHO (World Health Organization) and FAO (Food and Agriculture Organization), which are responsible for international scientific programmes and research institutes. Many of the formal institutions and informal associations, such as NATO and the OECD, that are associated with science internationally really reflect the corresponding groupings in the political, economic, and military dimensions. But most of these intergovernmental bodies are managerially and financially weak in comparison with the larger multinational commercial firms, which are also major elements in the system.

The complexity of international relations makes this pluralism unavoidable. Diplomacy, national security, trade and finance are important factors in the planning and performance of scientific activity, and are not themselves coordinated at the international level. Indeed, experience with both national governments and intergovernmental organizations suggests that the benefits of such coordination might not outweigh the costs of bureaucratic rigidities, timidities and delays.

Multinational science is growing and changing so fast that it has not settled down into a fixed institutional pattern. The international environment has so many extra dimensions that the relative costs and achievements of different organizational models are extremely difficult to assess. Experience is still taking political and administrative authorities up the steep part of the learning curve in such matters.

8.8 Coordination, collaboration or concentration?

In many fields of science and technology, a major part of the R&D is now organized multinationally. But this does not always mean that it is managed by a multinational organization. Within Europe, for example, most of the money that goes into 'collaborative research' is really spent on programmes where the work is only loosely 'coordinated', across national frontiers. In the official terminology of the EC, this means 'joint definition of aims, avoidance of unnecessary duplication, organization of meetings of research teams and exchanges of scientists, and wide dissemination of research results'. In other words, it demands little more of researchers and research entities than the active practice of the ethos of cosmopolitanism.

In principle, a programme of linked projects carried out in parallel by separate research teams is a good way of tackling a major scientific problem from a number of different points of view [§5.8]. In practice, the success of such a programme depends on getting the participants to cooperate actively and on putting in sufficient resources to cover the quite considerable costs of international communication and travel. Real progress on significant problems now often requires that research entities should 'collaborate' – that is, agree on a common research objective and carefully divide between them the labour of working towards it. In high energy physics or space science, for example, this inevitably means that the separate groups should combine to form a single large team. Scientists and technical staff in several countries have to be in daily contact, as they work together on various specialized functions of an elaborate piece of equipment. Even in fields such as geology or ecology, where the research action is more dispersed, collaborative research requires careful theoretical planning, standardized methods, and reliable performance by all participants. In effect, the condition for successful multinational scientific collaboration is the creation of a multinational research entity, coherently managed and directed.

Genuine international collaboration in science can thus be very time-consuming and expensive to arrange and undertake. It is not a process than can be initiated and steered from the top down, by administrators and managers. The working scientists, especially the leaders of the relevant research groups, have to be drawn in from the beginning, and they have to want it to happen. This process is bound to be highly selective for quality. In the absence of an overarching managerial structure, this selectivity necessarily involves some form of voluntary peer review, which is not easy to set up across national frontiers.

The diversity of national styles in the organization of R&D is another factor favouring rather loose and devolved administrative arrangements at the international level. Thus, for example, the French R&D system is highly centralized, although laboratory heads actually play tense games of bureaucratic in-fighting at the middle levels, and elected representatives of scientific employees have a surprising amount of influence. In Germany, by contrast, the central organs of science policy are weak, and are mainly concerned with sharing out resources fairly among the professors, each of whom has great power over the more junior staff in his or her 'chair'.

Coordination schemes must also be flexible enough to allow for unexpected scientific developments, and for the normal effects of quality control and peer review. Balancing up the relative levels of resourcing in different countries can present problems. British scientists, for example, could not be expected to collaborate with a French or German group with much higher standards of equipment [§3.2], especially if they suspected that they were being conned into a scheme by which the UK government was progressively reducing its overall contribution to 'public' science.

Countries also vary in the way they define distinctive areas of science, and distribute them amongst various sectors of the national R&D system [§4.5]. These differences are so salient at the higher levels of organization that formal agreements to collaborate internationally often have little practical substance. It often turns out to be much easier to match up scientific interests and responsibilities between two countries at a relatively low level. International collaboration in science has to begin with research entities who already share certain research traditions and who are motivated to come together voluntarily by common technical interests [§8.2].

Nevertheless, the question remains whether this sort of enlightened self-interest is sufficient to generate and hold together the communal scientific institutions now required on a global scale. Now that science in most countries has 'gone national', many of the scientists and scientific administrators whose work has expanded more and more into the international domain are beginning to doubt whether the concept of national sovereignty is useful, or even meaningful, in the world of science. Even if they are not employed directly by an international organization such as CERN or UNESCO, they feel that a transnational human enterprise such as science ought to be under transnational control, and directed towards transnational goals.

This ideal is already embodied in a certain number of international institutes [§8.3]. All the problems of 'coordination' and 'collaboration' are resolved by 'concentration' – that is, by gathering researchers from a number of countries on to a single site, organizing them into multinational teams, and putting the whole outfit under international management [§6.8]. In effect, such institutions reproduce at the international level all the features of a typical national, regional or local research institute. Indeed, many 'national' scientific institutions have as strong transnational linkages, and are almost as cosmopolitan in their scientific personnel, as those that do not legally belong to any one country.

It is not obvious, however, that the concentration of researchers and resources into specialized international institutes is the shape of things to come. In most fields of academic science, the essential critical mass for effective research can be achieved through international 'networks of excellence' loosely linking numerous smaller centres [§3.6]. In spite of many political glitches, the world-wide effort now going on to map the human genome is a much better model for international collaboration in science than

EMBL, or even CERN. Apart from activities such as space science or plasma physics, which are closely connected with immense technological enterprises, the administrative framework scarcely exists to manage research effectively 'on behalf of mankind'. Indeed, the concept of a comprehensive international research system dealing with all the ills of humanity leads to all the follies, farces and horrors of research planning in a command economy [§5.9]. It completely underestimates the dynamical tensions and intellectual ferments by which 'steady state' science must be kept open for enterprising, risk-taking individuals and institutions.

Whatever structures do evolve to manage research internationally, they will have to be compatible with 'steady state' conditions. Thus, for example, the scientific and technical directors of institutions and research programmes will have to be accountable to a lay authority, not only as managers but also as prime agents in setting research objectives [§5.3, §9.3]. The question of who sets the agendas, and who are the effective gatekeepers for policy, will have to be clearly answered [§5.2]. In many cases at present, it is not clear whether control really lies in the hands of an international bureau, or whether the most powerful individual funders – governmental or commercial – have the final say.

The 'steady state' criteria of scientific excellence and/or social relevance are even more difficult to apply to international scientific institutions than they are to large national research establishments. Such institutions acquire immense inertia – whether in maintaining their momentum towards their original scientific objectives or in passively resisting revitalization and change.

The point is not that science is being taken over by the bureaucrats of Brussels, Whitehall, Washington, Swindon, Bethesda or wherever your particular demon kings are based [§6.6]. On the contrary, for political reasons, it seems extremely difficult within an intergovernmental organization to exercise adequate control over the quality of research or researchers. The officially prescribed evaluation procedures [§5.5] cannot be applied with sufficient rigour to exclude weak participants, and distant political authorities find it particularly difficult to assess and influence the technical and administrative policies of an entrenched scientific elite.

An important role thus devolves on international bodies such as the OECD, the ESF and the ICSU, which are scientifically well informed, and politically well connected, but which do not themselves support research on a large scale [§8.6]. They are thus in a position to evaluate national and international scientific programmes and institutions from a much more independent point of view than organizations such as the European Community, where structure and action are intertwined.

8.9 National science policy in an international context

Every country in the world – including the United States – is beginning to realize that it can no longer go it alone in research. Scientific activity in other countries, whether or not it is organized multinationally, has a direct effect on the national R&D system. In a European country such as the UK, the national scientific enterprise is already an integral part of European science.

This realization has radical implications for science policy. For example, one of the main motivating factors in the transition to 'steady state' science – international competition for technological advantage in industry, weaponry, etc. – has to make a U-turn towards international cooperation. For the UK, this change is fully justified politically by the need to cooperate industrially with other European countries, in trade competition with Japan and the United States. Nevertheless, it generates considerable structural change within UK science.

The diffusion of the 'steady state' rationale from the national to the international scene colours all aspects of science and technology policy. Thoughts turn, for example, to the possible benefits of international specialization and the division of scientific labour [§3.9] within the European Community. Governments know that they must contribute their fair share of cash or kind to the total research effort in such communal sciences as medicine, hydrology or criminology [§8.4]. They have a very strong incentive to minimize the duplication of such efforts by subscribing to an overall plan. Organized selectivity, in the sense of national specialization by research fields, seems an almost inevitable development [§6.8].

Indeed, it is quite clear that no single European nation – perhaps no nation in the world – can now expect to be at the forefront in all fields of science. Intense competition for resources and esteem leads almost automatically to a tacit division of labour. Quite apart from any deliberate attempt to force the pace of specialization, this is generated continually by the normal working of the intellectual marketplace, as the banner of esteem in any one subject passes from centre to centre and from country to country.

Experience at the national level suggests, however, that these natural forces now operate too slowly and erratically to match the rapidly increasing scale and complexity of competitive research in science and technology. At the moment, every government in the world professes to value its scientific autonomy just as highly as the other traditional attributes of national sovereignty. New countries – for example, the independent states emerging out of the former Soviet Union – solemnly survey their national capabilities and potentialities in research, and draw up national plans for science and technology [§5.7]. Unfortunately, every one of these national plans argues for concentration of effort in almost exactly the same areas – 'new materials', 'biotechnology', 'supercomputers', and so on. This means that every country is adopting almost the same order of priorities, and tends to think of itself as on its own, all the way back through the 'pre-competitive' stages of research to the most basic scientific activity.

The problem, then, is how to break the symmetry of mutual competition and move into a mutually cooperative mode of action. It is absurd, of course, to dream of a comprehensive multilateral agreement to coordinate a whole discipline or sub-discipline by allocating specific domains to specific countries. In any case, research fields cannot be handed out internationally like the provinces assigned to the various powers at the Congress of Vienna, without concern for the human interests and technical capabilities of its existing inhabitants. This does not mean trying vainly to protect a national tradition of research in a particular field when the people and the promise have gone away. It simply means that international selectivity has to grow naturally out of national policies of building institutional strength around groups of people who have already demonstrated their ability to work effectively in given research areas.

There is general agreement that even a relatively small country, such as the Netherlands, Denmark, Switzerland, or Austria, can still find all the resources needed to maintain a few institutions containing a number of world-class research entities. But they have had to come to terms with the fact that their scientific capabilities are very patchy. The result of increasing specialization must be that even a major scientific nation such as the UK ends up without any centres of excellence at all in one or more of the main domains of science. This is a very disturbing prospect, since it could leave the country without the capacity to exploit a foreign scientific discovery offering valuable technological, commercial or military opportunities. Thus, for example, the quite unexpected discovery of high-temperature superconductors in Switzerland caused a great deal of concern in many other countries where research on superconductivity had been abandoned for lack of any foreseeable industrial applications.

In a highly competitive technological world, it is very advantageous to have early, privileged, even secret, access to research results capable of practical application [§2.7]. It is very difficult to compensate for this advantage by subsequent financial transactions, such as royalty payments for patents. In the past, this would only apply to fields of research that were already associated with commercially profitable technologies. But the applications of basic science have become so multifarious and diverse that the risk of missing an exploitable opportunity now extends over fields of science where no commercial benefits are at present foreseen [§2.3].

For this reason, international specialization in research is a two-edged policy. It promises economies through the reduction of duplicated efforts; but it also seems very risky to agree not to work in a field that might just possibly yield unforeseen benefits in due course [§6.8]. Policymakers argue – and are, of course, supported by the scientists whose work is threatened by such a policy – that it is essential not to reduce national research capabilities in any field to such a low level that they could not be expanded fairly rapidly to meet a new situation. The conventional wisdom is that international scientific collaboration in a given area must not go so far as to threaten the 'defensive threshold', whereby the country can still have a 'watching brief', or a 'gatekeeper presence', implying sufficient investment in manpower and facilities to replicate and improve on foreign advances as and when they occur.

This sort of calculation is reasonable enough for an advanced industrial country with a high-quality science base. But what about the situation for a developing country whose meagre educational and economic resources effectively debar it from reaching international standards in any major field of science. It is unfortunate that many well-intentioned bilateral and international efforts to overcome these disadvantages have proved very vulnerable to financial instabilities, or to ill-informed insistence that research should be directly related to visible national needs [§5.7].

The role of 'steady state' science as an instrument of national economic development may also engender the suspicion that the scientific transactions between two countries may not be equally beneficial to both parties. For example, an agreement to collaborate in pharmaceutical research may, in effect, be little more than a licence to exploit the citizens of a country with weak legal controls over mass clinical trials. Even in the relationships between scientifically sophisticated countries such as the UK and the USA, the financial power of the major partner determines how the research is programmed and who profits from its results.

International factors are thus moving rapidly from the margin to the centre of national science policy. Even in scientifically powerful countries such as the UK, these factors are reinforcing all the other 'steady state' pressures on the research system. They are both hastening the transition of science into a new regime and making this regime more unmanageable. At the international level, there is even less linkage than nationally between the public and private sectors, between the research supported by intergovernmental organizations and the R&D undertaken by multinational corporations. The tension is growing between two contrary desiderata: on the one hand the openness required for international collaboration in basic research; on the other hand, the secrecy engendered by heightened international competition in technological R&D.

8.10 Unifying the scientific world

Since the early 1970s, a significant proportion of academic science has thus been 'internationalized' in quite a new sense. Individuals and research groups are not simply members of a world community, competing internationally for recognition for their contributions to a world-wide knowledge base: many of them are now tied administratively or financially into the projects and programmes of transnational bodies, for whom they carry out research as employees, subsidiaries or clients [§8.5]. Scientists are not just communicating voluntarily with their colleagues in other countries and travelling abroad from time to time to meet them: they are moving house and taking up regular employment abroad in order to collaborate with them in multinational teams. They become accustomed to working for weeks, months, or years abroad, often among colleagues from a variety of other national cultures.

The mass exodus of research workers from the former Soviet Union and its former satellite countries shows that there is now a world-wide market for skilled scientific labour. Competent scientists quickly learn that their capabilities and formal qualifications fit them for employment in any well-found research institution, in any country [§7.13]. They are naturally attracted towards those centres where they can combine good science with good living conditions. Their specialised education [§7.12] – usually gained at the considerable cost of their native country – often eases them through visa and work permit formalities. They transform the cosmopolitan ideal [§8.1] into a professional reality.

Governments, of course, see the voluntary emigration of 'their' scientific personnel rather differently. An adverse 'brain drain', is now a major policy concern for many countries. This is actually a very complex social phenomenon, which is not necessarily deplorable. Even the obvious direct economic effects – the loss of the local wealth-creating potential of a talented, expensively trained, indigenous group of people – may sometimes be out-weighed by the indirect, long-term benefits of world-wide scient-ific and technological progress. It is hard to imagine, for example, how New Zealand would have gained if Ernest Rutherford had not made his scientific career in England, but had returned in 1910, say, to his old university to teach rather elementary physics. To strike a more sombre note: Germany, along with the rest of the world, would have been vastly poorer now if Albert Einstein, Max Born, Hans Krebs, and many other outstanding Jewish sci-entists had stayed in their own country in the 1930s, and perished in the Holocaust.

Personal migration from the periphery to the centre has always been a feature of Western science. Calculations of individual advantage have always motivated scientists to travel from all over the world to the Meccas of research, whether these were in Europe before 1939, or in the United States since 1945. Indeed, one of the main purposes of European scientific collaboration since that time has been to create centres, such as CERN, that would reverse this traffic. The evolution of the Community towards a 'single Europe' is designed to eliminate all the barriers to personal mobility throughout the region, whether for short visits or for permanent employment. In effect, the whole Contin-ent is supposed to become, like the United States, a single 'aca-demic marketplace', where scientists will compete freely for research posts, according to their talents and interests, without regard to nationality [§7.3].

Quite apart from the general political obstacles to its realiza-tion, this federalist scenario is still Utopian. Laboratory life itself is cosmopolitan, but scientists, like other middle-class citizens, are tied into their national cultures by the mundane strings of language, education, leisure activities, families, working spouses, friendships, etc. A move to Naples, say, for an inhabitant of

Amsterdam still looks far less attractive, and far more risky, than, say, a move to New Orleans for someone from Boston. In effect, it means becoming an expatriate. People do make such moves, but seldom frequently or easily.

This vision of a unified scientific world in which scientists are personally free to move and compete for jobs and esteem is the Holy Grail of cosmopolitan individualism. This is the aspect of international scientific cooperation that attracts the leaders of scientific communities in every country. But the systematic international coordination of research under political auspices is opposed, for the same reasons that individual entrepreneurs oppose the coordination of industry. It has implications that threaten their personal independence, their expertise their commitments – and even their livelihood.

In their eyes, the growth of transnational scientific traffic should not be regarded as a move towards a unitary research system under the control of supernational quasigovernmental authorities. Like the growth of transnational financial transactions, it should be seen as a means of enhancing the competitive forces favouring efficiency and quality of performance. Research entities should continue to be constituted and supported – perhaps even by their parent academic institutions or by sub-national regional bodies – as if they were indeed independent small firms, offering alternative products to discriminating customers [§5.10]. International action should be limited to the provision of indivisible research facilities, or the improvement of the communication infrastructure, whether by scientific publishing, electronic linkages, or personal travel [§3.6]. Indeed, if the economists are to be believed, any external interference with the strategic and technical autonomy of each research entity should be considered a restraint on trade, leading to sub-optimal outcomes in terms of input/output efficiency or product quality.

As we have seen, transnational science does follow this 'market' model in many ways. International competition, as measured, for example, by the publicity surrounding the award of Nobel Prizes, is certainly fiercer than ever; yet the individuality of research entities and academic institutions is carefully respected [§7.9]. Research entities are treated as free agents when they coordinate their individual research plans in large-scale environmental programmes; the analogy is with the small engineering firms that supply prescribed components to motor or aircraft manufacturers. A multinational 'Big Science' facility such as CERN is less like a supermarket than it is like one of those cooperative wholesale organizations that purchase goods in bulk and distribute them to small shops.

But the pole of perfect competition is as far from current realities as the pole of perfect coordination. Taken to its logical extreme, the market analogy leads back to the traditional situation where individual scientists were supposed to be quite free to formulate and carry out their own research projects, subject only the approval of a highly specialized group of peers. It is doubtful whether there ever was such a golden age for scientific enterprise, and we are certainly not in such an age now. The public authorities that now provide patronage on a vast scale for basic science not only insist that the expenditure of their money should be strictly accounted for: they also demand a voice in the specific purposes for which this money is to be used.

The notion of the international science base as a 'level playing-field' is a mirage. Nowadays, the competitive strength of a research entity in the international marketplace depends as much on the funding policy of its government as on its own innate scientific excellence. National funding bodies are simply not willing to underwrite the inevitable wastage of a system of outright scientific *laissez faire*, and introduce various elements of 'coordination' to economize on their science budgets. Scientists seem to dream of something like a global science foundation that would fund research projects, big and small, in a 'responsive mode', where only technical excellence and human need would score [§5.8]. There is no reason to suppose that an international body of this kind would be any more generous, any less politically or commercially influenced, than the National Science Foundation (NSF) or the NIH in the United States, the Science and Engineering Research Council (SERC) in Britain, or the DFG in Germany.

The position we have reached was not obvious at the beginning of this chapter. The rhetoric surrounding the academic ethos seemed to say it all: that science had always transcended national frontiers, and was becoming more global than ever [§8.1]. But this is a plea for cosmopolitanism, not for internationalism in the organizational sense. The 'Big Science' rationale for the latter is quite clear, and is certainly expanding to cover a larger and larger proportion of the research effort, even in basic and strategic science. The globalization of R&D in multinational companies is also a very significant recent phenomenon which will soon interact with academic science at the same level.

Nevertheless, the essential tension in the transnational arena is still between 'coordination' and 'competition', between the 'command' system and the 'market' system, just as it is in national science policies [§5.9]. Both these systems favour cosmopolitanism, but for quite different reasons. International coordination requires the scientist to play a more active organizational role in an international structure: transnational competition requires him or her to trade more actively in a transnational market. In this aspect, as in others, 'steady state' science is certainly not going to be a staid and static enterprise, sheltered from a stormy world in a remote ivory tower.

9
Steering through the buzzword blizzard

So the greedy peasant took a sharp knife and cut open the goose that laid the golden eggs.

9.1 A game with moving goal posts

We began this book with the visible evidence that science was passing through a period of radical structural change. Subsequent chapters presented a deeper analysis of the origins and extent of this change, showing in detail how it affects every aspect of scientific life and work. To give substance to the analysis, I have focussed on the UK, where a transition to quite a new regime started in the mid-1970s, and is still going on. But the history of UK science in recent years is only part of a larger story, involving scientific activity throughout the world. Science has always been a multinational cultural form. The same forces for change are at work everywhere, and many of the same features have emerged in other countries.

The notion that a distinctive new regime is emerging may be too simple. In some countries, the transformation process is apparently only beginning, and may take a different turn. Everywhere, the scientific world still seems like the Irishman's impression of New York: 'Sure, it'll be a fine city when it's finished.' On the other hand, scientific activity is becoming increasingly transnational, both for general political and economic reasons and in response to the pressures of the 'steady state'. UK science, for example, is now inextricably caught up in the complex of interconnected institutions that is developing in Europe. National research systems, like architectural styles and economic policies, tend drearily to follow international fashions.

The story of change is evidently very far from complete. We are still in the midst of a major historical event, whose contours and outcome we can only guess. There is no way back to the traditional way of managing the business of research, but there is also no obvious path forward to a cultural plateau of comparable stability. The new structures that are emerging are not the products of a gentle process of evolution: they are being shaped very roughly by a dynamic balance between external forces exerted by society at large and internal pressures intrinsic to science itself.

These forces are too powerful to be modified or deflected. In most cases, the best we can do is to pit them against each other, and learn to live in the region of turmoil where they meet. The trouble with 'steady state' conditions is not that they are static: it is that they are altogether too turbulent for comfort or efficiency. A striking characteristic of many of the present features of UK science is that they are essentially provisional. Faced with novel demands and situations, the responsible authorities at every level have been improvising wildly. They feel that they are in a game where they must score immediately, before the goal posts are moved again. The whole system has become extraordinarily fluid. Nobody is quite sure what arrangements will crystallize out and harden into a regular pattern of principles, procedures, policies and practices for the longer run.

9.2 New and old requirements

As politicians say when they have no idea what to do, this is a time of challenge and opportunity. The challenge is to understand properly what is now happening to science, and to think imaginatively a few years ahead. The opportunity lies in the many decisions that will have to be taken as we move into a new era. The immediate issues are so salient – and so complex – that they

tend to dominate the scene. But the way that particular issues
are visualised and resolved has a lasting influence. Real strategic
thinking manifests itself less through formal plans than through
innumerable tactical decisions, each directed towards a clearly
perceived objective.

Unfortunately, this objective cannot be defined as a coherent
whole. The scientific enterprise is much too complex, interactive,
diverse, inconsistent, self-actuating and contextually unpredictable to be represented by a working model and then redesigned
to a new set of blueprints. It is sometimes a useful exercise to
work out plausible scenarios, where desirable but contrary features are cunningly balanced, but these are usually pipedreams.

Realistically, the most that we can usually do is to try to understand how an effective research system actually functions, so as
to make sure that *essential* functions are not impeded as a result
of seemingly harmless organizational change. A great deal
depends not on *what* is done, but on the *way* that it is done. As
we all know to our cost, it is in the small print that so much of
value is unwittingly given away.

The difficulty is that the research systems in many countries
are being refashioned so rapidly that it is not easy to determine
how they *do* function, except by virtue of the momentum imparted
to them under more favourable conditions. The various practices
through which the essential functions for 'good science' were
traditionally performed have been melded with or replaced by a
whole host of new practices and procedures, each designed to
meet the needs of a passing moment. In many cases, it is not
even clear whether these functions are being carried out at all,
even though they are vital to the long-term health of the
enterprise.

This certainly does not mean that all the novel features of
'steady state' science are entirely opportunistic, or that they
threaten the ultimate viability of the scientific endeavour. There
can be no dodging explicit demands for competitive excellence,
economic use of resources, managerial efficiency, or the systematic exploitation of scientific discoveries. These, after all, are perfectly reasonable requirements in themselves, especially when
they come from the body that ultimately provides most of the

money: a democratically elected government. They cannot be disregarded just because some of the means that have been rigged up to impose them are wasteful, insensitive, unduly bureaucratic or otherwise burdensome.

Nevertheless, excessive zeal in the interpretation of these requirements has given birth to practices that are quite out of keeping with the research process. Many of these practices are so ill judged that they could do lasting damage to the health of science and its efficacy as a social institution. Let us look at some of the things that are now going on – things that would certainly be immediately apparent to the hypothetical time-traveller whose observations were imagined in chapter 1.

9.3 'Accountability'

Take the requirement of 'accountability'. What this means literally is that scientists should be able to show that they have used the resources they have received for research in accordance with the terms on which those resources were provided [§5.2, §5.3]. It implies demonstrable trustworthiness in carrying out a task for which one is responsible and answerable. No scientist could possibly object to such a condition of employment or funding – provided that it was fairly phrased and interpreted.

Nobody, for example, would defend the diversion of research funds to high living, lavish hospitality, or other forms of personal or institutional gratification. Again, breaches of the ordinary norms of personal integrity, such as dishonest research claims, are intolerable, not only because they defraud the taxpayer but also because they tear great holes in the delicate web of mutual trust that supports the whole research process. And yet such pathologies do occur. The most notorious recent examples come from the United States, where some very distinguished universities have been challenged dramatically over the indirect costs chargeable to research grants [§6.5], and where there have been several long-running sagas involving accusations of fraud in the entourage of some very eminent scientists [§5.3].

Modern science is organized on too large a scale to deal with such follies and crimes informally, behind the scenes, without recourse to financial audits and quasi-judicial investigations. But experience shows that it is impossible to specify, detect, prove and punish every deviation of this kind. The danger is that public indignation, political opportunism and administrative diligence will combine to impose very detailed and rigid financial accountancy on research funding, and quite unrealistic modes of scrutiny on research performance. As the scientific community is uneasily aware, this aspect of 'steady state' science is all too vulnerable to the damaging effect of political attack, which could saddle it with very heavy bureaucratic burdens.

The important point, however, is that the concept of 'accountability' covers much more than detailed financial and intellectual probity. It implies positive efforts to achieve the objectives for which resources have been provided, and the avoidance of unnecessary waste along the way [§6.4]. In other words, it indicates that these resources should not merely be devoted to the agreed purpose, but that they should be seen to have been used 'efficiently', as judged retrospectively by the donor.

One of the characteristics of 'steady state' science is that most of its funds originate from high-level governmental or commercial bodies. Such bodies are continually urged to apply hard-nosed input/output criteria of efficiency to all their operations and to ensure that these are focussed on clearly defined goals [§5.7]. When funds are short they naturally tend to assess scientific activity in the same terms. This attitude is encouraged by the practice of allocating resources competitively to proposals for research projects that promise to achieve highly desirable practical ends. Considerable expertise is needed to interpret such promises realistically. Without such expertise, it is all too tempting to treat these conjectural objectives as if they were genuine outputs, and to assess each project as if it were a simple exercise in getting 'value for money'. This, in turn, encourages researchers to overemphasize the (hypothetical) material profitability of their projects, rather than what they might contribute to knowledge.

The requirement of 'accountability' thus lays stress on the narrowly instrumental aspects of science, at the expense of its exploratory, speculative aspects [§5.6]. Applied unimaginatively, it would punish a deviation to exploit a serendipitous opportunity that had not been allowed for in the original research proposal. Carried to extremes, this attitude could be quite devastating, for it would starve most of the science base, and lay short-term 'value added' conditions on all other research. There is even a tinge of populist suspicion that scientists following their own personal research trails, in their own time, at their own expense, might be up to no good, and ought to be made 'accountable' to society for their activities.

Even when interpreted very moderately, this demand is slowly pushing 'steady state' science into more instrumental modes. Thus, for example, many of the agencies set up or greatly expanded by a number of governments in the 1950s and 1960s to support various branches of academic science – for example, the UK Research Council system, the US National Science Foundation, and the Australian Commonwealth Scientific and Industrial Research Organization – have been quietly restructured or recommissioned to contribute much more directly and explicitly to 'national economic competitiveness' or even to earn their keep through industrial contracts. The economic rationale of such developments is impeccable – except that it is quite blind to the unpredictable, incalculable, but equally genuine benefits that are sure to stem in the long run from much less 'accountable' research.

Peer review procedures tend to be unfavourable to genuine scientific originality. It is often remarked, for example, that neither Darwin nor Einstein would have received alpha gradings for their unconventional research plans. Nobody really believes that Newton might have produced much more in his middle years if he had been forced to submit his inconclusive alchemical investigations to the scrutiny of his peers. Insistence on 'accountability' is essentially contrary to the personal trait of 'enterprise' that has always been among the prime characteristics of scientific excellence.

9.4 'Evaluation'

In the wake of 'accountability' comes 'evaluation' [§5.5]. The efficiency of a business cannot be estimated without knowing what sums to enter in the earnings columns of the ledgers. Can you put a value on a particular piece of research – or on a particular researcher? This is not regarded nowadays as an absurd question. Scientific leaders in the US have been greatly cheered by careful calculations showing that the economic return on federally funded research is not far off 30 per cent per annum. Information gained by research is regularly bought and sold as 'intellectual property' [§2.7]. There is controversy over whether accountants are justified in treating an accumulated stock of research results as an 'intangible asset', comparable in value to the sums 'invested' in R&D by a company over the years. The research performance of each 'cost centre' (i.e. academic department) in every UK university is assessed and assigned a numerical indicator which determines arithmetically the grant it will receive [§6.1].

For obvious reasons, the actual figures that come out of such exercises are not to be taken seriously outside their peculiar organizational contexts. But they are no more than logical extensions of the requirement that all research projects, programmes, entities, and institutions should be 'assessed' [§5.4], that all researchers should be personally 'appraised' [§7.7], and that 'indicators' should be determined and published for all aspects of research performance. To the outside observer, these multifarious manifestations of a general doctrine of 'evaluation' must surely be one of the most salient features of 'steady state' science.

Again, what objection could there be to this in principle? Research is such a specialized activity, with such imponderable products, that it cannot be taken entirely at its own valuation. Too much is now at stake to allow science to go its way without systematic, detailed assessment of the quality of what is done [§5.4]. The real question is just how this essential function should be carried out in practice [§6.8].

Consider, for example, the spacing of 'evaluations' in time. Hard-pressed funders and managers are tempted to demonstrate their *machismo* by carrying these out much too frequently. Research has a variety of rhythms, ranging from the year or so that might be required to produce a publishable research result to the decades of purposeful effort that have often gone into the making of a major scientific revolution. Individuals, research teams and institutions undergo characteristic cycles of renewal and decay, each with its own time constant [§7.13]. A detailed 'evaluation' cannot fail to interfere seriously with the dynamics of the cycle into which it is inserted, if only by the diversion of effort into putting a good public face on work that one knows privately to be perfectly sound. A succession of such events within the same cycle has all the effects of frequently pulling up a seedling by the roots to see how well it is growing.

Despite being contrary to the most elementary worldly wisdom, this neurotic syndrome is already endemic in 'steady state' science. Self-winding principal investigators, who would measure the evolution of their scientific careers in five- or ten-year phases, are expected to undergo annual Job Appraisal Reviews [§7.14]. Triennial 'research performance indicator' exercises are imposed on university departments whose scientific quality – and academic staff – could only be changed through a decade or more of dedicated effort [§6.1]. New programmes or units are promised five years of undisturbed funding, and then 'reviewed' after two years under threat of premature liquidation.

Experienced research managers know that very close scrutiny of research performance is counter-productive. Only mediocre researchers doing unoriginal work require frequent, systematic, unremitting surveillance. High-quality research is not, superficially, a highly efficient activity. Its performance cannot be judged by people who have little experience and no significant achievement in a research environment. Of course there are other well-tested recipes for crushing scientific creativity, such as overlavish funding, control by ancient has-beens, poor training, or harsh constraints on travel, not to mention career uncertainties, quite inadequate resources, or demands for instant results. But there is a very real danger that 'steady state' science will fail to obtain all the excellence it expects from researchers and research institutions just because that excellence is demanded altogether too insistently.

An even more serious feature of many 'evaluations' is that they are carried out incompetently. The rationale of all judgement is that the judge is operating from a higher level than the subject of judgement. This is peculiarly difficult in science, where significant actions are both ill-defined and narrowly framed. A full evaluation of the professional performance of a research scientist is a formidable task in itself, and cannot be entrusted to a single referee. The only persons who can do this properly are other scientists specializing in the same field. The list of these is always very short, and they can usually claim that they have better things to do. 'Peer review' [§5.4] is an indispensable element of any 'evaluation' process in science, but is exceedingly expensive of top-level scientific expertise.

Aficionados of scientometrics claim that they have a technological fix for the labour intensity of 'evaluation'. But quantitative indicators which have some significance for the comparison of very large data sets – for example, national scientific productivity in major disciplines – are unconvincing on their own for, say, the appraisal of individuals or small research entities [§5.6]. In science, there is *no* substitute for expert human judgement. A research system that makes undue use of formal 'evaluation' procedures is either wasting the means it has for producing good science – better science, often, than the work being evaluated – or else obfuscating itself with masses of perfunctory or inept assessments of its own activities.

There are worrying signs that 'steady state' pressures are pushing some national research systems towards one or another of these conditions. The UK science base is already suffering severely from the second malady, self-inflicted by its own funding bodies. The three successive 'Research Performance Indicator' exercises [§6.1, §6.8] have all been far too large and elaborate for the human resources that were available to carry them out. As a result, academic scientists have lost all confidence in the basic validity of the procedures that were used. They see their research careers being shaped by 'evaluation' procedures which they widely regard as hollow, maladroit, and sometimes insensitive to the point of rank injustice [§7.14]. They can be forgiven the cynicism with which they then attempt to boost their ratings by flooding the data banks with low-quality papers or mutually favourable citations.

This is not the place to argue about whether or not these crude procedures actually get more or less 'the right answer' (whatever that means!). The fact that they are so widely criticized and distrusted represents a grave challenge to the legitimacy of authority in the research system. Working scientists actually spend a great part of their time trying to estimate the value of their own or other people's efforts [§7.3]. In time they develop an exquisite sense of what constitutes genuine quality in an examinee, a graduate student, a collaborator or competitor, a candidate for a post or a prize, a manuscript, a published paper, or a research proposal. When the occasion demands, they are accustomed to putting much thought and labour into a considered assessment of scientific quality – even though they know that such assessments vary inexplicably from assessor to assessor. It strikes to the heart of their professional commitment, and their respect for well-earned scientific reputation, to make so much depend on the outcome of a hurried, slapdash process which takes so little direct account of what they hold to be the true values of their calling.

9.5 'Selectivity'

In the wake of 'evaluation' comes 'selectivity'. The history of science often gives the impression that all scientific progress can be attributed to discoveries due to a small number of outstanding researchers. Although this elitist view undervalues the contributions of the many, and over-values the good luck of the few, it contains an important truth. Scientific ability is very unevenly distributed. The function of 'academic freedom' is not to facilitate typical scientific careers, or to optimize the research productivity of typical researchers: it is to leave openings for a few untypical people with untypical talents.

'Selectivity' makes good sense if it is based on a thorough and valid assessment. Research is not an egalitarian profession. It is a rigorous pursuit, where incompetent performance, as signalled by persistently low achievement, eventually clogs up the system. Under 'steady state' conditions, there is no margin that can be frittered away on uncompetitive people or projects. Firm procedures of various kinds are required to channel resources to the best researchers and research entities, and to take away resources from those that cannot use them well. The pruning metaphor – 'we must cut out the dead wood to allow the living tissue to flower and bear fruit' – is harsh but appropriate.

As we have seen [§6.8], the competitive spirit of academic science generates hard choices whose cutting edges are sharpened under 'steady state' conditions. But 'selectivity' means more nowadays than choosing the best candidates for patronage or preferment and allowing research groups whose proposals do not win alpha ratings to fade away. It is associated with a policy of deliberately building up certain groups and actively killing off others for reasons that are not directly related to proven scientific quality.

These decisions usually refer to a notion of 'viability', encapsulated in managerial clichés about the need for 'concentration', in order to build up a 'centre of excellence' with a 'critical mass' of research effort. The idea is to give an artificial boost to the collectivization processes that are natural to the transition to 'steady state' conditions. Fascinated by the phenomenon of Big Science, scientific leaders and managers have come to believe that the *size* of a research entity is a significant measure of its scientific potential, regardless of other structural factors [§3.6]. In particular, they often assert that a research group smaller than a certain size cannot hope to be effective, and that it should either be expanded forthwith, or else disbanded so as to free its members for more productive employment.

This belief is not well founded. Expected economies of scale and scope are often negated by human factors. The effectiveness of a research group depends less on the number of its members than on their individual capabilities and their relationships with one another. Neither 'contributions' nor 'recognition' satisfy the laws of arithmetic. They vary enormously in their intrinsic value and are not additive quantities. Under these circumstances, an average over a large group is extremely misleading, since it tells us nothing about most of the population except that they must be well below this average. The sheer visibility of a large unit or institution often gives an impression of overall quality that does not survive detailed examination. The excellence of a new centre cannot be bought with lavish funds and prestigious appointments: it has to be won by achievement over a period of years.

This is not to repudiate the strategy of giving strong support to what is really good in science, and withdrawing support from what is mediocre. But the demand for 'selectivity' frequently leads to arbitrary decisions to close down small, but beautiful, research operations, stifle novel developments that have not yet made big waves, and disregard talented individuals who happen to work in marginal institutions. In practice, it is often just a code word for the application of the Matthew Effect [§3.9, §5.6]. By deliberately favouring those who are already excessively favoured, undue 'selectivity' legitimates, reinforces and enhances the existing hierarchies of size and status in the research world. Procedures designed to loosen up the scientific enterprise by removing unproductive components may actually make it more rigid.

9.6 'Manpower'

Science is the occupation of large numbers of professionally expert, well-motivated people. Under 'steady state' conditions, it is important that such people should be available when required. National research systems thus concern themselves about future supplies of research personnel, and make policy statements about how this vital requirement is being met. In practice, official forecasts of supply and demand in the scientific labour market are never borne out, but this causes no harm since the corresponding policies are never carried out.

Nevertheless, the notion that 'steady state' science requires a supply of 'manpower' is not as innocent as it seems. I am not referring to the customary use of this masculine terminology, despite the substantial and growing proportion of women in the research profession. This usage is deplorably insensitive, but the fact that it has not been superseded by some gender-neutral term, such as 'personnel', probably has little significance. That is to say, the transition to 'steady state' science does not seem to have been particularly favourable to the careers of women scientists, but it has not, as such, increased the barriers and prejudices against them [§7.14].

One of the implications of this term is that people need to be 'trained' to do research. This, again, is perfectly harmless if it is interpreted broadly to cover all that would-be scientists undergo, or do, to gain the various skills they need for their highly specialized work. As we have seen [§7.12], this includes a long period of formal education, followed by doctoral studies and post-doctoral experience. It is a lengthy process, punctuated only occasionally by the acquisition of formal qualifications such as the academic degrees of Bachelor, Master, or Doctor. More significantly, in the later stages it is largely a process of *self*-education, driven by increasing confidence in one's powers as a 'self-winding' personality.

In circles outside science, however, the notion of 'research training' is interpreted much more narrowly and instrumentally. It is taken to refer solely to the period of post-graduate study, leading to the award of a PhD, and treats this as a 'process', in which research scientists with appropriate skills are 'produced' as required. Pressure is then applied to improve the 'efficiency' of this process, as if it were not very different from the training of accountants or engine drivers.

This conception of graduate studies is clearly revealed in various clumsy attempts to reshape them to fit the supposed needs of 'steady state' science. Treasury officials send minutes to education ministry officials, drawing attention to the waste of resources implicit in low 'completion rates' – that is, the failure of many doctoral students to present their theses for examination at the end of the three years for which they are supported by government funds – regardless of whether that gives time enough for significant research experience. Universities are encouraged to give all scientists and engineers instruction in business management [§7.10], to prepare them for organizational niches in the enterprise culture, regardless of the intentions of many of them to stay in research. Doctoral programmes are scrutinized to ensure that students are given some training in all the research methods currently fashionable in their discipline, regardless of whether these methods are likely to be needed in their actual research.

So far, demands such as these have not struck seriously at the heart of the PhD experience, which is the psychological transition from a state of being instructed on what is already known to a state of personally discovering things that were not previously known. It is true [§7.12] that there is wide debate in the academic world on how best this transition should be initiated, supervised, and made to bear fruit. What is dangerous about the notion of 'research training' is that it could eventually stifle this essential experience under layers of other activities designed simply to produce well-informed people with useful practical skills.

The concept of a 'trained research worker' tacitly devalues the sense of personal commitment that motivates scientists throughout their careers [§7.8]. It seems to class them with people such as plumbers, barristers or orchestral musicians, who are employed to do very highly specialized jobs – but jobs that are defined for them by other people. They are made to feel like hired hands, who may indeed be worth the money they earn, but can be fired if they fail to do good work or if their services are not required. Their notion of their own creative autonomy is denigrated as a false ideology.

As a result, the deplorable social phenomenon of the middle-aged 'contract researcher' has become an established feature of 'steady state' science [§7.9]. In 'manpower' terms, this practice seems entirely logical: if appropriately skilled personnel are available to carry out a series of short projects, then economic efficiency requires that they should be employed for correspondingly short terms. Experienced and thoughtful academic scientists have been induced to accept this as a regrettable necessity, even though they know very well that it is an oppressive, anxiety-ridden way of life and a crazy, fragmented way of doing research. Here again, an apparently reasonable requirement is carried to a quite unreasonable extreme.

The conflict between the corporate interests of academic institutions and the personal interests of their academic staff is not just the usual power struggle between 'the management' and 'the workers' in an increasingly authoritarian structure. It arises from their involvement in different but overlapping market systems, whose diverse trading arrangements have not been systematically harmonized. This discord arises at many points: project funding procedures, internal institutional budgeting and management, conditions of employment, career prospects in teaching and/or research, and the assignment of intellectual property rights. As the environment becomes more competitive, these contradictions become ever more serious and damaging to science.

9.7 'Exploitation'

Science is largely funded nowadays on the promise of long-term material benefits. These benefits can only be obtained if research knowledge is eventually 'exploited' – that is, incorporated into a new product, process or practice. Under 'steady state' conditions, it is no longer acceptable for academic research to cut itself off from applied science and technological development. The trickle of potentially 'exploitable' knowledge out of the science base has to be turned into a steady flow, available on tap wherever it might be needed in national life [§5.7].

This requirement is so reasonable that one wonders why it should have to be emphasized so vehemently, and given such a high policy priority in most advanced countries. As we have seen, the frontiers between 'science' and 'technology' are scarcely visible on the ground [§2.3], and interpenetrate so promiscuously that no amount of ethnic cleansing could now separate them. In any case, one would have thought that working 'technologists' would always be on the lookout for useful information and ideas, and that working 'scientists' would always be delighted at the prospect of putting their discoveries to practical use. It is a mystery why they should have to be exhorted, and even bullied, to come closer together.

A possible solution to this mystery is that the issue is fundamentally misconceived. The fault lies in the 'linear model' of technological innovation, which grossly over-simplifies this extraordinarily complex social process. The fact is that the results of academic research seldom emerge with legible clues as to their practical applications – if any. History shows that these applications, when they occur at all, are often very far away from what might initially have been inferred [§2.4]. A useful artefact normally combines information, ideas and techniques garnered from innumerable scientific and technological sources, often in quite different disciplines or industries. By its very nature, this is a very elaborate and lengthy process. It can only be speeded up – as war-time experience showed, and as Japanese firms have further demonstrated – by throwing into the battle whole armies of money and skilled personnel. There does not seem to be any motivational or organizational gimmick that will do this on the cheap. In other words, loud talk about the need for much better 'exploitation' of science should not be taken literally. It is often designed simply to divert attention from failure to put enough public or private money into strategic and applied research, higher education, or corporate R&D.

Nevertheless, in spite of its shaky theoretical foundations, emphasis on organizational links between academia and industry is characteristic of 'steady state' science [§7.11]. The result is a great deal of institutional reform and innovation, some of which may well have the intended effects and much of which is beneficial for other reasons. It works with the grain of techno-scientific progress, encouraging the establishment of research entities and research programmes in areas that straddle the traditional boundaries between academic and industrial problems [§6.7]. It encourages personal mobility by providing temporary stopping points along career paths between the two sectors. It looks like the wave of the future.

Once again, there are dangers if 'exploitation' mechanisms are given absolute priority throughout the research system [§5.8]. 'Steady state' conditions do not require, for example, that all institutions of higher education should be closely connected with commercial firms, in the hope that this will ensure that their research results are put to use. Private sector organizations have their own agendas, which do not include the technological development of every wild idea that comes their way. Far from facilitating 'exploitation', a close linkage hinders the dissemination of information about work that might have unsuspected applications elsewhere. Some academic research entities may indeed thrive on unadventurous, short-sighted science, focussed on near-market technological problems passed on to them by industrial R&D managers. But it would be disastrous if this policy were to take over the whole science base, whose principal function is to perform open-ended research with more distant, more exciting prospects.

Quite apart from administrative constraints on the publication of research results, large R&D organizations are not tolerant of the public expression of 'negative' opinions. Technical controversy may be strongly encouraged internally up to a certain level in the administrative hierarchy, but orthodoxy (as defined managerially) is expected to prevail in relations with top management and with the outside world.

Commercial patronage of university research is encouraged by all governments because it replaces hard-pressed public expenditure [§4.2]. But, apart from the monetarist prejudice against government spending as such, it is not easy to see a deeper purpose in diverting private sector funds and managerial effort into forms of university research from which they are most unlikely to profit directly. This is especially true of the UK, where industry is notoriously sluggish in expanding its own R&D investment to match overall economic growth.

It is fashionable nowadays to regard the conflation of academia and industry as a primary virtue and a salient indicator of institutional excellence. But academic institutions have special responsibilities, such as post-graduate and post-doctoral education

[§7.12], which do not fit them well for research on urgent practical problems. The pace and security of research are much more difficult to control in a university than they are in a company laboratory. A shotgun marriage between such different cultures may produce offspring that are much less intellectually or technologically fertile than either of their parents.

The doctrine of instant 'exploitation' also forces the pace in the delicate negotiations that are often required when researchers from different disciplines arrange to collaborate [§3.8]. For example, various modes of collaboration between neurophysiologists, psychologists and computer scientists have produced remarkable progress in our understanding of how the brain functions. This work will undoubtedly produce many practical benefits; but the process of establishing a new transdisciplinary field would have been jeopardized if every collaborative project had to include from the start researchers whose only interest was how to apply the results. The paradoxical effect of such a demand might be that new scientific disciplines will have even greater difficulty in emerging and gaining a finger hold in academia than in the past. This could be particularly serious for some aspects of the social sciences, where easy access to new interdisciplinary areas is far more necessary than in the natural sciences.

9.8 'Priorities'

In the top strata of 'steady state' science, the buzzword is 'priorities'. It evokes the hard thinking that precedes the agonizing choices that have to be made in cutting a cake of limited resources to feed a hungry multitude of worthy proposals. It conveys an image of a carefully compiled list, ordered according to equitable criteria on the basis of objective information obtained by systematic procedures. It implies relative rather than absolute judgements, thus turning the anger of unsuccessful applicants away from the decision- makers towards those who have been successful. Excuse the clichés: this is one of the most useful words in the policy lexicon.

In effect, 'priorities' are the stuff of serious, top-down policy-making [§5.7]. There is no way of doing this rationally except by stating various objectives, working out how they might be attained, deciding their relative importance, and assigning resources accordingly. Of course there are other ways in which various scientific activities actually get funded, such as the whim of a top person, a power struggle between autonomous agencies, the local interests of elected politicians in their pickings from a pork barrel [§5.2], or the sentimental responses of donors to heart-plucking charitable appeals. But such practices are grossly offensive to the proclaimed rationality of science itself, however acceptable they may be in other political spheres.

The trouble is that the goals to be prioritized and the means of reaching them are extremely uncertain [§5.6]. Looking down into this fog, the politicians, officials and industrial managers who must set the priorities can make out only the broadest outlines of the ultimate benefits of research in a particular area. The danger is that this innate uncertainty will be forgotten as decisions become more tense. Conflicts and anxieties are resolved by specifying areas of emphasis in more and more detail. Vague hopes and aspirations are firmed up into expectations and promises. Diligence in developing a coherent national *policy* for science eventually generates documents that could easily be mistaken for that political fantasy – a national research plan [§5.7].

Everybody knows, surely, that the production of scientific knowledge is more risky, more diversified, and more interdependent than even, say, the production of consumer goods for an advanced society. Intelligent people do not really believe that it is possible to collect and process the information and understanding that would be needed to guide operations at every laboratory bench, or even in the office of every laboratory director, or even at the meetings of research councils and university senates [§5.9]. Nevertheless, a quite unusual degree of overt *dirigisme* has been an obvious feature of the transition to 'steady state' conditions.

Listing 'priorities' [§5.7] is a relatively harmless way of raising consciousness of science in political circles. Most such lists actually refer to technological R&D, which has commercial firms and other powerful corporate interests to defend it. But the science base is more vulnerable to efforts to plan it in detail from the top, even though such efforts are nearly always ineffective. Generally speaking, they do little direct damage, since scientists and scientific institutions learn how to circumvent ill-informed directives, or else they produce 'plans' for projects whose outcome they can foresee with confidence because they have already carried them out. But the system gets clogged up with dishonest and/or unrealistic project proposals, research programmes and corporate plans which cannot be properly assessed and are usually wisely ignored.

The real danger of extreme concern for 'priorities' is its top-down attitude to the production of knowledge. It reinforces managerial authority in the tree tops at the expense of expert initiative at the grass roots. It gives clout to the requirement of 'accountability' by setting out in detail the objectives to be achieved. Not merely must researchers show that they have interpreted responsibly the terms on which they are funded: they must also accept whatever terms their funders set, including the precise goals and methods of their investigations [§7.9].

This shows up, for example, in the shift of resources from 'responsive mode' funding to 'directed' or 'programmed' research [§5.8]. In other words, instead of selecting scientifically meritorious projects from a wide range of unsolicited proposals with very diverse objectives, funding bodies select particular problem areas, specify in some detail the objectives that they would like to see achieved, and advertise for project proposals promising to work towards those objectives. In this way, the funding bodies try to direct the front line of the science base into the areas that stand highest in their list of 'priorities', as distinct from the areas that researchers themselves see as the most open for scientific advance.

Even when research programmes are developed in consultation with leading researchers, they still constrain the range of questions to which research is to be addressed. Directed research has no place for the maverick project, challenging a policy norm, an academic fashion, or a supposedly established research result [§5.12]. In the natural sciences and their related technologies, the scope for intellectual conformism is limited because everybody knows that the obstinate realities of material fact will eventually break through. Somebody will do the experiment that disproves the theory; the plane will crash; the patient will die. But in the social sciences, the lessons of experience are harder to interpret and longer to take to heart. Under such circumstances, it is particularly foolish to rely on policy-related research that is programmed in accordance with the 'priorities' of the organization that sponsors it. 'Steady state' conditions do not favour support for high-quality social scientists doing such research in an institutional context that protects them from these pressures. Emphasis on 'priorities' is thus a very serious threat to the integrity and credibility of the social sciences, and the service they do to society.

9.9 'Competition'

It is the conventional wisdom of our age that the *market* system is superior to the *command* system. In straightforward politico-economic settings, where it is a matter, say, of getting consumer goods into the shops, this is not now seriously to be disputed. But this wisdom is not compelling at every node of the social world. Open markets for research claims and scientific reputation are far and away the best mechanisms for getting reliable knowledge out of the science base [§5.11]; that does not mean that 'competition' should be the rule throughout the research system.

We may, perhaps, overlook the notorious expenditure of time and effort involved in staging elaborate open competitions for resources [§5.3, §6.2]. Inevitably, far more research proposals have to be prepared and assessed than could possibly ever get funded. Market theorists would say that this wasted labour, infuriating as it is for the unsuccessful applicants, is a necessary cost of the quality improvements that competition generates. They would also say that even greater hidden costs are incurred through covert personal rivalries inside 'command' systems.

What is more serious is that extreme competition may actually be counter-productive in its effects on performance [§5.12, §6.5]. It is all very well to insist that individuals or institutions will be 'animated by market forces' and will go all out for excellence of the desired kind. In many cases, instead of a positive reaction, the immediate response might rather be described as stalling tactics, or playing safe, or aggressive defence of vested interests, or simply cutting back for the sake of a quiet life. In other cases, this response can be quite erratic, as people are pulled to and fro by intense but inconsistent forces, such as competition for personal reputation, competition for commercial profits, and responsibilities towards their disciplines and institutions [§7.6]. If pushed to excess, financial stringency can be wholly destruct-ive. It may cause the excellent Dr X to pull out of research, the promising Dr Y to leave the country, and the eminent Dr Z to take to drink!

In principle, competition should engender innovative pro-posals. In practice, it probably inhibits flexibility in responding to a serendipitous opportunity. It is very difficult for a small research entity, with all its personnel tied up in on-going projects, to muster up the very modest resources required to follow up quickly a chance discovery. Indeed, so far from encouraging 'enterprise', small-scale economic uncertainty can be a recipe for scientific caution. The leader of a research entity on the verge of 'insolvency' can scarcely be blamed for seeking commissions for absurdly short and unadventurous investigations that will at least keep together an expert team. Peer reviewers and specialist panels are often supposed to be unsympathetic to imaginative, unconventional project proposals: it may be that such projects are simply not coming up for consideration because the 'financial' return on them is considered too distant, and too risky, by the research entities who might be proposing them.

The basic principle of all market systems is that they are driven by an 'invisible hand' which optimizes the outcome of innumer-able separate transactions. The question remains whether sur-vival in this struggle necessarily represents scientific quality. Evolutionary fitness only has meaning relative to a particular environment. If that environment openly favours attributes such as 'accountability' or 'exploitability', then proposals seeming to have those attributes will flourish. In the time-honoured tradition of 'grantsmanship' [§5.6, §7.9], applicants then exaggerate these aspects of their projects, as they try to adapt their work to the ecologically viable niches in the system. Specialized expertise is then needed to penetrate the presentational gloss, and prevent research drifting in directions that are administratively convenient or politically opportune, rather than scientifically well conceived [§5.10].

Market forces work very imperfectly unless they operate in an atmosphere of accessible, trustworthy information about all aspects of the commodities being bargained for. Genuine competition in science is always by design quality and performance – which are enormously variable – rather than by price [§7.3]. And yet purely economic considerations suggest that the selection between superficially similar research projects should be based on their relative costs, rather than on the relative capabilities of those who propose them. Research units may adapt to enhanced 'competition' by trying to do too much, too quickly, to the detriment of the intrinsic quality of their work. It is not only that trying to do research on the cheap often turns out to be an expensive mistake: it is that cost is a very weak indicator of intrinsic worth in this trade.

In reality, the notion that 'competition' rules is far from OK. On the contrary, there are whole stretches of communal life where *cooperation* is paramount [§8.8]. Indeed, in most markets the transactions are not between individuals, but between organized groups, such as wholesalers, supermarket chains, government departments, universities, banks, manufacturing firms, etc. These groups require their members (i.e. employees, staff, partners, *et al.*) to work together towards common goals. The same cooperative behaviour is an essential element in less formal groups, such as families, clubs, sports teams, voluntary associations, and even political parties.

When scientists say that they belong to a community, they are not just trying to cover up the scandalous truth that they often behave badly towards one another as rivals for esteem. In many ways, science is a single vast enterprise, held together at all levels by voluntary cooperation in a variety of essential activities. Quite apart from the team work and other forms of intellectual collaboration which are now required to carry out most research projects [§3.7], there are numerous communal institutions to maintain, such as learned societies, publishing houses, editorial boards, peer review panels, academic appointment committees, and even the governing councils of many of the bodies that fund the science base.

Excessive emphasis on 'competition' undermines the communal spirit that motivates people to take part in these activities, or transforms it in into a strictly self-centred desire to make a personal profit out of them. Under 'steady state' conditions, where money talks louder than esteem, communal functions and roles are commercialized, and expected to pay. Every move in the research game – consulting a data bank, using a clinical specimen, citing a reference, publishing a paper, providing a referee's report, photocopying a journal article, advising a government committee, giving expert testimony in a court of law – is treated as a market transaction, for which cash has to flow.

It is not so much that many of these moves are only notionally competitive, and are not really associated with marketable commodities. It is that, once money is involved, it is much more difficult to maintain that people are acting without calculation of personal advantage. The scientific enterprise – especially the science base – runs on trust, which depends on reasonable conformity with the norm of 'disinterestedness' in all public aspects of the scientific life [§7.6]. This norm is not compatible with commercial practices at vital nodes of the system.

9.10 'Management'

If there is a single word that epitomizes the transition to 'steady state' science, it must be 'management'. And yet this simply means that the groups, units, centres and institutions that perform research must be properly organized and led [§6.6]. Academic resistance to the very idea merely reflects regret at the collectivization of what was once a highly individualistic activity [§3.9].

Beneath the nostalgia, however, there is a more significant concern. Is science getting the style of 'management' that is appropriate to its needs and functions [§5.8]? It is all too easy to accept the logic of a convergence between the 'academic' and 'industrial' modes of research [§7.10]. Both modes now obtain their resources from large bureaucratic structures, they both have to be managed according to similar principles of accountability, efficiency, monitored performance, etc. – and they both must foster scientific originality, individual initiative and critical independence of mind. The 'academic' objective of designing and building an enormous new particle accelerator would seem to require the same managerial and administrative capabilities as the commercial objective of designing and building an enormous new power reactor or telecommunications system. The commercial objectives of a pharmaceutical company require an environment encouraging the same type of personal creativity in the discovery of new drugs as would be needed in a university laboratory for the discovery of new physiological mechanisms. Ergo, they should both be managed along the same lines.

Unfortunately, this does not answer the question. Closer examination of industrial R&D [§6.6] reveals that it is managed in very different ways in different industries, and in different firms within the same industry. Just as in academic research laboratories, one can find the whole range of organizational structures and management styles, from the quasi-egalitarian to the ruthlessly hierarchical – and every style has its successful and unsuccessful examplars.

What most industrial R&D organizations have in common, however, is that they must interface reasonably closely with their parent bodies, typically much larger commercial corporations run on standard bureaucratic principles. This creates limitations, demands and opportunities which do not apply to a more basic research entity embedded in an academic institution or governmental agency. Industrial R&D does not, therefore, provide well-tested prescriptions for the 'management' of the science base, even under extreme 'steady state' conditions.

In practice, the various organizational requirements packaged under the term 'management' have been imposed so rapidly that research units and institutions have not had time to draw on the actual experience gained with various possible systems in various settings. They have usually put in place elaborate structures that seem to conform with general opinion on 'how these things are done' – that is, with imitations of models that were generally regarded as the best practice in large commercial firms a generation earlier. And of course, in accordance with the French maxim, these arrangements being provisional, they have endured.

As a result, the stereotype of 'management' that has become established in 'steady state' science is of a heavy, top-down, bureaucracy, continually trying to stir its stumps towards unattainable objectives, but bogged down in sloughs of information that it cannot process. This stereotype is very damaging, since it is quite clearly inappropriate for almost all the actual activities supposedly being managed [§6.6]. It is also essentially incompatible with the academic culture of devolution, consultation, consensual decision, and personal and departmental autonomy.

9.11 Fundamental principles for the advancement of knowledge

'Steady state' science has come to stay. It is not yet a stable, fully formed system, but it will have to satisfy a number of new requirements. These requirements are all perfectly reasonable in principle, but they open doors to some extraordinarily foolish practices. These practices are not vital, as such, to the health of science – indeed some of them can only be described as pathological. It is the responsibility of scientists and officials, politicians and business executives to detect or foresee, eliminate and prevent, these pathologies before they become established in the system.

After this depressing litany of follies, let me finish on a more constructive note. Reflecting for a moment on the past history of science, one can identify certain features that contributed to its success. What are the *essential* requirements of the research process? What fundamental principles should the architects of 'steady state' science keep firmly in mind as they mould and remould the old structures to fit the needs of a post-modern world?

The real problem is to disentangle these principles from the network of social practices in which they are embedded. Are such cherished features of science as refereed journals and tenured academic posts fundamental to its operations, or are they simply possible ways in which certain vital principles can be made operational? What is really being sacrificed as we move out of the shelter of traditional practices?

One might have thought that answers to these questions would have emerged long ago from the busy world of academic science studies. Unfortunately, the 'metascientists' – the historians, philosophers, sociologists, psychologists, economists and political scientists who describe and analyse science and technology from a variety of different points of view – have not yet come up with a coherent account of just how the research process actually works. Indeed, much of their contemporary discourse seems aimed at proving that there is no such process, that anyway it doesn't work, and moreover that it works all too well on behalf of certain sinister power groups. I am not saying that such sceptical, or cynical interpretations are ill conceived or out of place: on the contrary, they have revealed many things about science and technology which have to be taken very seriously into account in the present study. But they do not offer direct guidance to the person faced continually with practical decisions, small and large, on how to keep the system going.

So the following list is presented very tentatively, as a starting point for much more extensive study and debate. But it is not just a 'programme for further research', designed both to postpone practical action and to provide work for some needy social scientists. This is not just an intellectual exercise: it is meant for real.

The list is brief, to keep the mind concentrated on essentials. It is very schematic, because each item could – and should – be the subject of a whole treatise. In any case, anyone who could not already compose a few paragraphs under each of these headings should not be allowed within reach of science policy! To the experienced and reflective practitioner, these particular items, expressed so briefly, may seem trite – mere motherhood and apple-pie values that everybody, surely, has at heart. Believe me, this is not the impression I get from some of the things that have happened recently, and threaten to happen, in the scientific world.

To come to the point. *Any research organization, requires generous measures of the following:*

- social *space* for personal initiative and creativity;
- *time* for ideas to grow to maturity;
- *openness* to debate and criticism;
- hospitality towards *novelty;* and
- respect for specialized *expertise.*

It is easy to think of further items, such as 'technical autonomy', 'stability of employment' or 'freedom to follow up serendipitous opportunities', but there is already beef enough in these five principles. They may sound too soft and old-fashioned to stand up against the cruel modern realities of administrative accountability and economic stringency. On the contrary, I believe that they are fundamental requirements for the continued advancement of scientific knowledge – and, of course, for its eventual social benefits. We have a very real responsibility to articulate them clearly for ourselves and for others, and to devise procedures, policies and practices that are consistent with them.

Further reading

This book is not the product of 'research'. It is not fashioned out of items of knowledge accumulated by systematic personal investigation, or by delving among the publications of other scholars. It is drawn directly from life. In fact the published and unpublished material relating to its theme is effectively infinite. Almost every issue of a general scientific or academic journal such as *Nature*, or the *Times Higher Education Supplement* contains a news item or comment that might be cited in support of some detail of the argument. Since the early 1980s, government departments, governmental agencies and legislative bodies in every country have been issuing policy statements, indicators, committee reports, corporate plans, etc., containing masses of other relevant information. The publications of international organizations such as OECD are excellent sources of comparative data: and so on.

This book does, however, have what sculptors call an 'armature'. It is built around a skeleton of concepts, theories, and generalized observations derived from a relatively small number of books, many of which can be recommended for further reading. The following list is not in any way a scholarly bibliography, but it provides entry points to the literature on some aspects of the particular topics discussed in each chapter.

PREFACE
Cozzens, S., Healey, P., Rip, A. & Ziman, J. (eds.) (1990) *The research system in transition* (Dordrecht: Kluwer)

1. WHAT IS HAPPENING TO SCIENCE?
Greenberg, D. (1969) *The politics of American science* (Harmondsworth: Penguin)

Price, D. de S. (1986) *Little science, Big Science . . . and beyond* (New York: Columbia University Press) (Reprint of edition of 1963).

Price, D.K. (1965) *The scientific estate.* (Cambridge, Mass.: Harvard University Press)

Ravetz, J. (1971) *Scientific knowledge and its social problems* (Oxford: Clarendon Press)

Salomon, J.J. (1973) *Science and politics* (London: Macmillan) (translated from French edition of 1970)

2. SCIENTIFIC AND TECHNOLOGICAL PROGRESS

Freeman, C., Clark, J. & Soete, L. (1982) *Unemployment and technical innovation: A study of long waves and economic development* (London: Frances Pinter)

Hagerstrand, T. (ed.) (1985) *The identification of progress in learning* (Cambridge University Press)

Nelkin, D. (1984) *Science as intellectual property: Who controls scientific research?* (Washington: American Association for the Advancement of Science)

Rescher, N. (1978) *Scientific progress* (Pittsburgh University Press)

Schäfer, W. (ed) (1983) *Finalization in science: The social orientation of scientific progress* (Dordrecht: D.Reidel)

3. SOPHISTICATION AND COLLECTIVIZATION

Weinberg, A. (1967) *Reflections on Big Science* (Oxford: Pergamon)

Whitley, R. (1984) *The intellectual and social organization of the sciences* (Oxford: Clarendon Press)

Ziman, J. (1984) *An introduction to science studies* (Cambridge University Press)

4. TRANSITION TO A NEW REGIME

Cozzens, S. & Gieryn, T. (eds.) (1990) *Theories of science in society* (Bloomington, IN: Indiana University Press)

Ezrahi, Y. (1990) *The descent of Icarus: Science and the transformation of contemporary democracy* (Cambridge, Mass.: Harvard University Press)

Rip, A. (1989) *Transformations of contemporary science* (De Boerderij: University of Twente)

5. ALLOCATION OF RESOURCES

Chubin, D. & Hackett, E. (1990) *Peerless science: Peer review and U.S. science policy* (Albany, NY: SUNY Press)

Franklin, M. (1988) *The community of science in Europe: Preconditions for research effectiveness in European Community Countries* (Aldershot: Gower)

Gummett, P. (1980) *Scientists in Whitehall* (Manchester University Press)

Irvine, J. & Martin, B. (1984) *Foresight in science: picking the winners* (London: Frances Pinter)

Kendrew, J. & Shelley, J. (eds.) (1983) *Priorities in research* (Amsterdam: Excerpta Medica)

Kohler, R. (1991) *Partners in science: Foundations and natural scientists 1900–1945* (Chicago: University of Chicago Press)

LaFollette, M. (ed.) (1982) *Quality in science* (Cambridge, Mass.: MIT Press)

6. INSTITUTIONAL RESPONSES TO CHANGE

Elias, N., Martins, H. & Whitley, R. (eds.) (1982) *Scientific establishments and hierarchies* (Dordrecht: D. Reidel)

Gibbons, M. & Wittrock, B. (eds.) (1985) *Science as a commodity: Threats to the open community of scholars* (Harlow, Essex: Longman)

Smith, B. & Karlesky, J. (1977) *The state of academic science: the universities in the nation's research effort* (New York: Change Magazine Press)

Whiston, T. & Geiger, R. (1992) *Research and higher education: The United Kingdom and the United States* (London: SRHE & Open University Press)

Wittrock, B. & Elzinga, A. (eds.) (1985) *The university research system: The public policies of the home of scientists* (Stockholm: Almqvist & Wiksell)

Zinberg, D. (ed.) (1991) *The changing university: How increased demand for scientists and technology is transforming academic institutions internationally* (Dordrecht: Kluwer)

7. SCIENTIFIC CAREERS

Bailyn, L. (1980) *Living with technology* (Cambridge, Mass.: MIT Press)

Ben-David, J. (1971) *The scientist's role in society: A comparative study* (Englewood Cliffs, NJ: Prentice Hall)

Caplow, T. & McGee, R. (1961) *The academic marketplace* (New York: Science Editions)

Hagstrom, W. (1965) *The scientific community* (New York: Basic Books)

Jagtenberg, T. (1983) *The social construction of science: A comparative study of goal direction, research evolution and legitimation* (Dordrecht: D. Reidel)

Latour, B. & Woolgar, S. (1979) *Laboratory life*: The social construction of scientific facts (London: Sage)

Merton, R. (1973) *The sociology of science* (Chicago: University of Chicago Press)

Ziman, J. (1987) *Knowing everything about nothing: Specialization and change in scientific careers* (Cambridge University Press)

8. SCIENCE WITHOUT FRONTIERS

Herman, R. (1986) *The European scientific community* (Harlow, Essex: Longman)

Hermann, A., Krige. J., Mersits, U. & Pestre, D. (1987, 1990) *History of CERN: Volumes I & II* (Amsterdam: Elsevier)

Index